GIANT APES: THE CULTURALLY DIVERSE CRYPTIDS

Author Name:
Alwaleed Alghanim

'Giant Apes: The Culturally Diverse Cryptids' Copyright © 2018 by Alwaleed Alghanim. All Rights Reserved.

All rights reserved. No part of this book may be reproduced in any form or by any electronic or mechanical means including information storage and retrieval systems, without permission in writing from the author. The only exception is by a reviewer, who may quote short excerpts in a review.

Cover designed by Cover Designer

Author Name: Alwaleed Alghanim
E-mail: alwaleedalghanim01@gmail.com

Printed in the United States of America

First Printing: Feb 2018

ISBN-13 978-1-9803210-3-3

You'll be amazed when I tell you that I'm sure that they (Bigfoot) exist.

—GEORGE NOORY

INDEX

Giant Apes: the Culturally Diverse Cryptids 1
Index 2
Preface 4
Introduction 6
Chapter I: The North American Bigfoot 9
    Section 1: Encounters 17
    Section 2: Evolution... 21
    Section 3: Theories 40
Chapter II: The Himalayan Yeti 60
    Section 1: Taxonomy... 65
    Section 2: Theories 67
Chapter III: The South American Mapinguari 78
    Section 1: Biodiversity 80
    Section 2: Ecology.. 83
    Section 3: Theories 86
Chapter IV: The African Waterbobbejan 95
    Section 1: Zoology.. 99
    Section 2: Theories 102
Chapter V: The European Wildman 110
    Section 1: Miocene Apes 117
    Section 2: Theories 120
Chapter VI: The Australian Yowie 129
    Section 1: Hominids... 134
    Section 2: Australian Hominids 142
    Section 3: Theories 145
Chapter VII: The Conclusive Theory 148
Conclusion 169
Acknowledgments 170
About the Author.. 171

GIANT APES: THE CULTURALLY DIVERSE CRYPTIDS

PREFACE

First and foremost, I would like to thank my family, friends, and the many professors that have assisted me for their help and encouragement in the creation of this book. This journey was not passed alone, but rather I passed it together with any others, especially with my parents, uncle, and really close friends.

Over the course of history, many different cultures have witnessed and theorized about a giant ape. From cavemen, to Norse Vikings, to Native Americans, people have long been debating over the authenticity of the witness accounts of the legends of Bigfoot and its relatives throughout the world. This, however, has increased dramatically in the 1800's, with the release of supposed videos and image evidence of the cryptid.

With intense debates revolving around the cryptid, especially with the North American Bigfoot, this book aims to display all of the witness accounts of the six major types of the legend and the scientific theories and pieces of scientific and mathematical evidence around them. The six major types of the legend have all been chosen as the most well-known types of the legend in each continent, with the exception of Antarctica.

Throughout the entire process of research for this book, many different perspectives have appeared on the plausibility of the cryptid. This caused many different theories around the cryptid, from the supernatural to the government conspiracies. With many perspectives, and thus theories, it is nearly impossible to state all of the theories and their plausibility. Thus, I have gathered several of the most popular and plausible theories about each type and discussed the probability of each theory. Thereafter, I have stated the most likely scientific explanation for each type. In the conclusion however, I have gathered all types of the cryptid and explained the several scientific phenomena that would most likely describe how people would have mistaken creatures known to science as the cryptid.

In a span of six months, I have gathered many research papers and mathematical equations that show intensive information that combine smoothly into a single, large theory. While the theory proposed in this book is not the final conclusion to the entire debate of the cryptids, the theory still holds strong pieces of evidence that would clear at least some of the most well-known misconceptions and theories about the cryptid.

INTRODUCTION

From the Bigfoot of North America and Almasty of Russia to the Yeti of the Himalayas and the Yowie of Australia, there is a name for the legend in nearly all cultures. For centuries, stories of a giant human-like ape have been terrorizing people throughout the globe. During a span of roughly two centuries, the timespan where the legend popularized and grew rapidly, there is yet any solid scientific evidence to prove the existence of the creature. The only scientific evidence taken seriously by both skeptics and believers is hair samples, which led to the biggest scientific Bigfoot project in 2013 by Bryan Sykes, a professor of genetics at Oxford University.

Throughout all cultures, the story of the beast is slightly different. In Russia, the Almasty is a hominid, mostly connected to Neanderthals. In the Himalayas, the Yeti is a surviving population of Gigantopithecus Blacki, the largest known primate yet to exist, supposedly going extinct 100,000 years ago. Bigfoot of North America is supposedly a giant unknown primate, perhaps a bipedal species of gorilla. Similarly to Bigfoot, Yowie of Australia is believed to be a species of primate. Whether it was the so-called Bigfootologists, scientists, or any person in that matter, has one question in mind: What type of animal are we looking at? Could it really be a giant undiscovered primate? Is it some sort of primitive human species that survived in some of the world's remote places? To believers, it's either one of the possibilities according to their cultural belief. To skeptics, they ask: Is it a misidentified animal? If so, what type of animal could this legend be? If not, then could it all be merely a hoax?

The search shouldn't be towards the creature itself, but for the answers for the questions that are asked when looking at the legends in a scientific viewpoint. Many supporters for the creature's existence state that science has rejected the study of the creature, and it is clear to see why. Claims and researches about scientists are typically lazy and easy to debunk, as though they have completely rejected the existence of Bigfoot. This completely defies the definition of science, as science does not accept or reject anything no matter how small and useless it seems. Unlike the impression given by the scientists, science looks at the facts and flows towards that direction. The term science is derived from the Latin word scire, which translates to 'to know' or 'understand'. According to Oxford Dictionaries, science is defined as:

'The intellectual and practical activity encompassing the systematic study of the structure and behavior of the physical and natural world through observation and experiment.'

In the western United States, there is a term called Bigfootology. A bigfootologist is a term given to a person who searches for the existence of Bigfoot. While the term 'Bigfootology' isn't a real term in science, there is a branch of science, which includes bigfootology: cryptozoology. Cryptozoology is the search and study of animals with an unsubstantiated existence, the most renowned of which are Bigfoot and the Loch Ness Monster.

Cryptozoology students learn about recent discoveries of animals, taxonomy and classification, research techniques, conducting field investigations, equipment, working with other researches, as well as the magnitude of the estimated number of undiscovered species. 'Bigfootology' is a specialized branch of cryptozoology called 'Hominology', which focuses on the study of primate cryptids such as Bigfoot as opposed to the study of all types of cryptids. In cryptozoology, there is a primary doctoral degree and two specializations. The primary doctoral degree is 'Doctor of Metaphysical Humanistic Science' while the two specializations are Cryptozoology and Hominology. The salary of cryptozoologists ranges between $36,000 and $93,000.

Professor Bryan Sykes, one of the world's leading geneticists, have obtained and examined some of the world's most convincing hair samples of Bigfoot from around the world. Looking at the donors of the hair samples in the documentary series, Bigfoot Files, there has only been one person to be experienced in this field: Nikolai Valuev. The problem, however, is that Nikolai Valuev was neither a donor nor a true cryptozoologist, which lowers his credibility.

With true hominologists being extremely rare compared to scientists of other fields, most notably in fields such as biology and astronomy, there are a few scientists willing to research about the legends. With only a handful of scientists involved, the study of Bigfoot is mainly composed of supporters that would easily misinterpret the given evidence. This would make any further research difficult, and finding the truth would be far more challenging.

There are countless versions of the infamous cryptid throughout the years, with many terrifying and persuasive records of Bigfoot with two famous examples being the duo Roger Patterson and Bob Gimlin, and Leif Erikson. It takes a lot of effort to explain such encounters, yet there are scientific answers to the mystery of the cryptid. From paleoanthropology, biology, and ecology to DNA tests and scientific experiments, there is an answer to the most widely known version of the legend in each continent. In North America, the beast is commonly known as Bigfoot, also referred to as the Sasquatch. In South America, it is known as the Mapinguari. In Europe, it is called the Wildman. In Asia, there are the Himalayan Yeti and the Russian Almasty. In Africa, the cryptid is known as the Waterbobbejan. Finally, Oceania's most well-known version of the legend is that of the Australian Yowie. As seen in the next several chapters, all of the stated examples can be scientifically explained.

CHAPTER I: THE NORTH AMERICAN BIGFOOT

Named after the explorer Americo Vesspucci, North America is the third largest continent. With an area spanning 24.71 million square kilometers, North America is the only continent that has all types of biomes. The highest point on the continent is Mount McKinley in Alaska, while the lowest point on the continent is Death Valley in California.

Mount McKinley is the former name of Mount Denali, which has a summit elevation of 6,190 meters above sea level. Roughly 4,250 kilometers away stands the lowest point in North America, Death Valley. Located in Eastern California, Death Valley has an area of 7,800 square miles and stands in 86 meters below the sea level.

There are eight types of biomes characterized by a particular climate, the quality of the soil, and plant life, with North America containing all. To clarify, biomes are defined as large geographic areas similar in climate and biodiversity. Starting from the north of the continent and going southwards, the biomes are as following: Polar ice, tundra, taiga, mountain zones, grasslands, temperate forests, deserts, and tropical forests.

Polar ice biomes are defined as the regions of the Earth covered by ice most of the years. An ice cap, which gives this biome its name, is high-latitude region of a planet or moon covered in ice. The term is somewhat inaccurately named since an ice cap has an area of less than 50,000 square kilometers, and is always over land. An ice sheet is a term given to a larger area of ice. Polar ice caps do have any size, composition, or geologic requirements of being over land. However, polar ice caps must be centered in the polar region.

Tundra is derived from the Finnish word **tunturia**, which means treeless plain. The tundra ecosystem is famous for its frost-molded landscapes, very low temperatures, little precipitation, meager nutrients, and shortly growing seasons. Decomposed organic material serves as a nutrient pool. There are two major nutrients in tundra, nitrogen and phosphorus. Nitrogen is created by biological fixation, while phosphorus is created by precipitation. Characteristics of tundra are an exceptionally cold climate, low biodiversity, a limitation of drainage, a short season of growth and reproduction, energy and nutrients found in the form of decomposing carcasses, and large population oscillations.

Taiga is a Russian term meaning "forest". During winter, the taiga is extremely cold with only snowfall. In summer, the taiga is warm, rainy, and humid. A lot of coniferous trees grow in the taiga, making it known as the boreal forest. Boreal is also the same name as the Greek goddess of the North Wind. Due to the harsh conditions in this biome, not many species live in the taiga. The taiga biome is generally susceptible to many wildfires due to the trees' adaptations by growing thick bark. The fires would burn away the upper canopy of the trees and let reach the ground. New plants will grow and provide food for the animals that once were not able to live there because there were only evergreen trees. Animals in the taiga tend to be predators such as the lynx and several members of the weasel family such as wolverines, bobcat, minks, and ermine. Herbivores in the taiga include snowshoe rabbits, red squirrels, voles, red deer, elk, and moose. The red deer, elk, and moose tend to be found in regions where more deciduous trees grow.

The Latin word for "High Mountain" is "alpes", which is where the Alpine biomes derive their names. Alpine biomes are found in the mountainous regions throughout the world, usually at an altitude of 3,048 meters (10,000 feet) or higher. The alpine biome lies just below the snow line of a mountain. In the entire height of a mountain, there would be several distinct biomes. For example, the North American Rocky Mountains begin with a desert biome, then the deciduous forest biome, the grassland biome, the steppe biome, and the taiga biome before reaching to the cold Alpine biome. During summer, the average temperature ranges from ten to fifteen degrees Centigrade, and dropping to below freezing during winter. Because of the severe climate of the Alpine biomes, species have developed unique abilities to survive such harsh conditions.

At high altitudes there is very little carbon dioxide, which plants need to carry on photosynthesis. Because of the cold and wind, most plants are small everlasting groundcover plants, which grow and reproduce slowly, protecting themselves from the cold and wind by hugging the ground. Alpine animals have to deal with two types of problems in the Alpine biomes:

The great quantity of high UV wavelength is caused due to having a less amount of atmosphere to filter the sun's UV rays. With the exception of insects, only warm-blooded animals are found in the Alpine biomes. Animals in the Alpine biomes adapt to the cold by hibernating, migrating to lower and warmer areas, or insulating their bodies with layers of body fat.

Grassland biomes are large, undulating terrains of grasses, flowers, and herbs. For the most part, latitude, soil, and local climates determine the types of plants that grow in particular grasslands. Grasslands are regions where the average annual precipitation is great enough to support grasses, and a few trees in some areas. Drought and wildfires prevent large forests from growing. Grasses can survive the wildfires because they grow from the bottom instead of the top, as trees do. Therefore, their steams can grow again after being burned off. The soil of most grassland is also too thin and dry for trees to survive. There are two types of grasslands: tall-grass and short-grass. Tall-grass are humid and very wet while short-grass are dry with hotter summers and colder winters than their tall counterpart. Both types of grasslands are found in the United States. In the United States, there are more than eighty species of terrestrial animals, three hundred species of birds, and hundreds of species of plants.

Temperate forests, also known as deciduous forests, have five different zones. The first zone is the Tree Stratum zone. In the Tree Stratum zone, there are trees such as oak, beech, maple, chestnut hickory, elm, basswood, linden, walnut, and sweet gum trees. The Tree Stratum Zone has a height that ranges between 18.3 and 30.5 meters (60 to 100 ft.). The second zone is the small tree and sapling zone, which has young and short trees. Shrub zone is the third out of the five zones. Some of the shrubs found in the shrub zone are rhododendrons, azaleas, mountain laurel, and huckleberries. The fourth zone is called the herb zone. This zone contains short plants such as herbal plants. Finally, the fifth zone is the ground zone. The ground zone contains lichen, club mosses, and true mosses.

Animals adapt to the climate of deciduous forests by hibernating in the winter and living off the land in the other three seasons. The animals have adapted to the land by trying the plants in the forest to see if they are well to eat for a satisfying supply of food. In the land, the trees provide shelter for the animals and are also used for food and water sources. Most of the animals found in the deciduous forests are camouflaged to look like the ground.

There are two types of deserts: hot and cold deserts. Hot deserts, as referred by their name, are hot and dry. Most hot deserts don't have a lot of plants, but some low down plants manage to survive on the harsh conditions. Due to the inability to live in the immense heat of the day, the only animals capable to survive in the hot deserts are the animals with the ability to burrow underground. These animals only come during the night where it is a bit cooler than during the day.

A cold desert is a type of desert that has snow in the winter, which opposes hot deserts that only drop a few degrees in temperature. In a cold desert, it never gets warm enough for plants to grow with the exception of a few grasses and mosses. Furthermore, the animals that live in cold deserts have to dig burrows in order to keep warm. Requiring the same ability to dig burrows underground is the reason why some animals exist in both hot and cold deserts.

Finally, a tropical rain forest is a forest of tall trees in a region that is warm throughout the year; with an average precipitation of 125 to 660 cm (50 to 260 inches) of rain yearly. Tropical rain forests now cover less than 6% of the Earth's land surface, covering less area than deserts, which are estimated to cover about 10% of the Earth's land surface. Scientists have estimated that more than half of all the world's living organisms live in the tropical rain forests, producing 40% of the Earth's oxygen. A tropical rain forest has more types of trees than any other region in the world. In South America, scientists have counted about 100 to 300 species in a 1-hectare (2.5 acres) area. Roughly 70% of the plants in the rain forest are trees.

About one quarter of all the medicines used come from rain forest plants. Curare comes from a tropical vine, and is used as both an anesthetic and to relax muscles during surgery. Malaria is treated by Quinine, which comes from the cinchona tree. A person with lymphocytic leukemia has a 99% chance that the disease will go into remission because of the rosy periwinkle. Estimates put over 1,400 varieties of tropical plants to be thought as a potential cure for cancer. Due to this, scientists might soon have a cure for the horrendous complex group of disorders. Now knowing the general geography of North America and the characteristics of each biome, further clues about Bigfoot can be found by looking at modern human history.

After the natives' civilization in the New World roughly 100,000 years ago, the first person to discover and colonize the New World is Christopher Columbus. Christopher Columbus was an Italian explorer, navigator, and colonizer. On the 12th of October 1492, Christopher Columbus and his crews aboard the *Nina*, *Pinta*, and *Santa Maria* ships landed in the Bahamas. Upon Columbus' return to Spain, news of his discoveries captivated Europe.

Although Columbus wasn't the first European to discover the Americas, with the title going to Norseman Viking Leif Erikson, Christopher Columbus' four voyages helped on opening trans-Atlantic navigation and eased the European conquest of the New World. Columbus made three subsequent journeys to the New World, locating many islands in the Caribbean and mapping the coasts of Central and South America.

Though Columbus refused to believe that he had discovered a new continent, others quickly grasped on the implications of his four voyages; with some people today believing that Columbus was hailed as a hero. Christopher Columbus even tried to force Queen Isabella to make good on her promises to appoint him as the governor of all his discovered lands, which led to a protracted legal case that still saw his successors in court over 150 years after Columbus' death in 1506.

Between the years 1519 and 1521, Hernán Cortés commanded a small Spanish expedition that would eventually capture the Aztec capital of Tenochtitlan, which would allow Cortés to take over as the governor of Mexico. Hernán Cortés was and is still known as arguably the most famous Spanish conquistador and the conqueror of the mighty Aztec Empire of Central America.

Hernán's willful disregard for authority pitted him against his superiors in Cuba and Spain, and died an embittered man in 1547 even with his exploits being legendary and quite profitable.

The conquest of the Aztecs from 1519 to 1521 by a small band of Spaniards and their native allies still ranks as one of the most formidable military feats in history. This is due to the fact that Hernán Cortés used tactics designed to divide the rival native nations of Mesoamerica in opposition of each other by using a large force of native associates to assist his attack against the loathed Aztecs in their island capital city of Tenochtitlan. Hernán Cortés was just as merciless with his own troops as he was with the citizens of Tenochtitlan, even going as far as to burn his own boats to ensure that his men would have no choice but to follow him. Cortés set the standard that upcoming conquistadores followed in the New World. It was through Cortés' ad Francisco Pizarro's efforts as well as others that the large Native American empires of the New World.

Author Name: Alwaleed Alghanim

Hernando de Soto was a Spanish conquistador who led a catastrophic expedition of conquest into the North American interior from 1539 up to his death in 1542. De Soto, hoping to follow in the footsteps of Cortés and Pizarro, failed in his hunt for gold on the North American mainland. After leading his men on a futile three-year search throughout of what is now known as the southeastern United States, de Soto died on the banks of the Mississippi River, and his men fled back to Mexico.

Although de Soto's journey was an unconditional failure, it was historically significant nonetheless. During his travels, de Soto encountered a densely populated and culturally advanced Native American civilization in the Mississippi Valley. Unintentionally, De Soto, his men, and his livestock introduced new diseases to the region that would eventually destroy the Native American populations. By the time Europeans returned to the region again in the late 17th century, the Mississippi Valley appeared to be a depopulated wilderness.

Moctezuma II, also known as Montezuma, was the last emperor if the Aztecs. In 1520, Moctezuma died in the captivity of Cortés. It is difficult to reach towards an accurate reading of Moctezuma's character and abilities because much of his history is unknown or known from extremely biased Spanish sources. What is known about Moctezuma is that the numerous people he conquered didn't love him. It is also known that Moctezuma's complicity with the Spaniards originally led to dispute within the Aztec nobility, which eventually led to his own death.

Moctezuma's demise was the climax of the Spanish story of the conquest of what is now known as Mexico. According to different sources, Moctezuma was either a simple ruler to some degree who was unable to comprehend the threat the Spanish presented to his empire, or a sharp-witted schemer who attempted to use the arrival of the Spanish to his advantage but was foiled by his own people. Moctezuma's death and the separation of his empire allowed the Spanish to firmly embed themselves in the New World. As a result, the riches of the Aztecs belonged to many conquistadores. Today, those who see him as the last native ruler of Mexico consider Moctezuma as a hero.

One of the most successful Spanish conquistadors is Francisco Pizarro. At the age of 57 in 1532, Pizarro led a small body of Spanish soldiers and seized the mighty Inca Empire. Pizarro later then founded the Spanish colony of Peru, ruling former Inca territories in that region up to his assassination by the followers of a rival conquistador in the year 1541.

GIANT APES: THE CULTURALLY DIVERSE CRYPTIDS

The ecological processes of the Columbian Exchange gave Pizarro an essential advantage of his conquest of the Inca Empire. The Columbian Exchange is a period of cultural and biological exchanges between the New and Old Worlds. Beginning from Columbus' discovery in 1492, the exchange lasted throughout the timespan of expansion and discovery. Developments in agricultural production, evolution of warfare, increased mortality rates, and education are only a few examples of the Columbian Exchange's effects on both Europeans and Native Americans. In 1525, a devastating smallpox outbreak killed nearly 250,000 Inca, including the emperor and many of his most powerful aides and generals, which led to a power struggle among the survivors that devolved into civil war. When Pizarro invaded the Inca Empire a few years later, he faced much less resistance than he would have prior to the epidemic.

Junipero Serra was a Franciscan friar who founded a concentration of missions in what is now known as California. At the age of seventeen, Serra joined the Order of Friars Minor in 1730 and asked to be sent to the Sierra Gorda Mountains of Mexico. Subsequently, Serra became the "president" of missions in Alta California. He helped in establishing nearly a dozen missions himself, and influenced those who established the remaining of the twenty-one California missions. In total, the missions stretched from San Diego to Sonoma.

Serra was the father of California's missions, and helped in establishing Spanish control over California coast during an era where Spain's authority over the area was under threat from England to Russia. Serra's missions were influential in converting all of the natives of California as far north as Sonoma into Christianity. The missions also constituted the first major European presence on the American Pacific Coast. California remained an important area of Spanish control until the start of the Mexican-American War in 1846, when it was captured by a group of Americans led by John C. Frémont.

During that era of discovery and expansion in the New World, there wasn't any Bigfoot encounter. Ever since Christopher Columbus entered the New World in 1492, it took 318 years for the first Bigfoot encounter in the New World to occur, which happened in 1811. 4 years after the first encounter in 1895, an article was written about Bigfoot, describing it as no more than a giant grizzly bear.

The grizzly was killed in Bald Rock, 96.6 km (60 miles) away from Fresno, California. California has one of the highest recorded Bigfoot sightings in America, but the history doesn't add up. How can such a giant beast remain undetected for over three centuries? Is it all just a hoax? Is Bigfoot actually a type of bear? To answer these, we must study the two most famous type of Bigfoot found in the continent: Bigfoot and Sasquatch.

To study Bigfoot, we must know the basic physical description of the famous cryptids by the most common piece of evidence of its existence, Bigfoot encounters.

SECTION 1: ENCOUNTERS

986 AD - New World (North America)
Leif Erikson and his men were in their first landing in the New World when huge, hairy men towered over them. According to the Norsemen, the manlike beasts were 'horribly ugly, hairy, swarthy, and with great black eyes'. Leif Erikson described them to live in the woods and had a deafening shriek and a rank odor. He and his men have described beasts that were clearly distinct from the native people. After several sightings of the huge creature, Leif Erikson and his men eventually departed the island, calling the beast 'Skellring', which roughly translates to 'barbarian'. It should be noted that the Norse were hairy people themselves, with matted hair and beards; and for them to consider the beast to be hairy, then the Skellring is much hairier than the Norse. This leads many people to believe that this is one of the earliest reported encounters of Bigfoot.

20/10/1967 - 1:15 - 1:40 PM - Bluff Creek, California
The most famous Bigfoot encounter is that of Roger Patterson and Robert "Bob" Gimlin, the two friends who have recorded the renowned Patterson-Gimlin film. In an early afternoon, Patterson and Gimlin were, in general terms, riding upstream in the northeast on horseback along the east bank of Bluff Creek. At anywhere from 1:15 and 1:40 PM, they "came to an overturned tree with a large root system at a turn in the creek, almost as a room." When Patterson and Gimlin rounded the overturned tree, "there was a logjam left over from the flood of '64". Nearly spontaneously, Patterson and Gimlin spotted the figure of Bigfoot behind the tree.

Bigfoot seemed to have either been crouching beside the creek to their left, or just standing there on the opposite bank. Patterson initially estimated its height to be anywhere around two meters tall, and later raised his estimate to 2.3 meters tall. Gimlin's estimation was a height of up to two meters. Some later analysts have suggested that Patterson's latter estimate was one foot (approximately 30.5 cm) too high, including the anthropologist

Grover Krantz. In the film, Bigfoot had either silvery brown, dark reddish-brown, or black hair covering most of its body. Overall, the figure seems to generally match the descriptions of Bigfoot given by those who have claimed to spot the cryptid. Patterson's pony reared up, and he grabbed his 16 mm camera, chased the creature, and filmed the entire scene. Meanwhile, Gimlin was sitting atop his pony with his rifle ready.

1811 - Jasper, Alberta, Canada

The first modern sighting of Bigfoot was in Canada, where it is known as the Sasquatch, 825 years after the sightings made by Leif Erikson and his men. In 1811 near what is now the town of Jasper, Alberta, Canada, a trader named David Thompson found some strange footprints. According to David, the footprints he found in the snow had four toes, and were roughly 36 cm (14 inches) long and 20 cm (8 inches) wide.

22/6/2009   6:30 PM   - Rhinebeck, New York

A 19-year-old college student was driving on a curvy back road on the way to a rehearsal in a nearby performing arts center. As the student swerved to miss what was later discovered to be a shopping bag containing an open cereal and a small log on the road, the student glanced in his rearview mirror. He saw someone or something darting behind his car, to seemingly retrieve the bag. After a moment, the student stopped and turned his car around and got a glimpse of a bipedal creature 15.24 meters (50 feet) away for 3 to 4 seconds. He described the creature that he spotted from the rear and side profile as a bipedal creature with a height of anywhere between 2.1 and 2.3 meters (7' and 7'6"), covered with black hair, and possessing muscular shoulders with arms that swayed in an exaggerated manner and palms that faced upwards. The student recalled that during the brief encounter, he felt "nervous, confused, and excited at the same time".

8/1/2008   1:30 AM   - Scipio, Utah

A big-rig driver was transporting a load of Idaho potatoes when the fog grew increasingly heavy as he downshifted and headed down an incline. The driver noticed something by the side of the road with glowing eye, and thought it might be a deer. When the driver switched on the high beams, he was startled to see a large creature running across the road in long strides around 6 meters (20 feet) away. The driver later estimated that whatever he saw was at least 2.4 to 3 meters (8 to 10 feet) tall and weighed anywhere between 272 and 362 kilograms (600 and 800 pounds). It had black hair, big eyebrows, and arms that were proportionately longer than those of a human. For a moment, the creature turned its head and stared at the quickly approaching truck. The driver swerved hard enough to avoid hitting the creature, nearly causing the truck to crash.

Fortunately, the driver regained control of his truck. However, as the driver managed to roll to a stop 183 to 247 meters (200 to 300 yards) away and look back at the creature, it was already gone. The driver, an avid outdoorsman and hunter, told an investigator from BFRO during an interview that he had always been skeptical about Bigfoot's existence. After the incident, the driver had changed his view, admitting that whatever he saw terrified him.

1/9/2009   6:15 AM - Rifle, Colorado

A woman commuting to her job was feeling a bit dazed on her way to work, despite her usual cup of takeout coffee. Just before the woman started up through the Independence Pass, she decided to pull her truck over to the side of the road and get a little bit of fresh air. As the woman got out of the truck, she noticed that something was moving in the meadow directly ahead of her. First, the woman thought it was a bear, but as the creature stood up, the woman noticed that its arms hung to its sides similarly to a person. The woman has stated to a BFRO investigator that the creature was huge with cinnamon-colored fur. After persuading the investigator, she also revealed that the creature had a pair of large breasts. Prior to the encounter, the woman admitted that she ever believed in the existence of Bigfoot and that her life has "forever changed".

1953 - 2017

A video was released in YouTube where a man explains that his father has shot and preserved a Bigfoot specimen in 1953. The man showed the head of the Bigfoot specimen, and claimed that it was around fifty kilograms.

The list of encounters similar to those stated is in the thousands, with 3,313 reports occurring in the 92 years between 1921 and 2013 in the United States by itself. Using statistics, estimates put the reports by 2020 at anywhere between 4,500 and 4,700 in the United States alone. With slight variation between each encounter in a vast area, it is known that Bigfoot is not one individual, but many that make up a single species. Nonetheless, there is a general description that might suit the healthy fully-grown adults of the entire species rather than one individual.

Bigfoot is described as being two to three meters tall, or 6'7" to 9'10", weighing in anywhere between 180 and 450 kilograms, or 396.8 to 992 pounds. Its footprint is 30 to 56 cm long, or 1 ft. to 1'10", 12 to 28 cm wide at the ball, or 4.7 to 11 inches, and 5 to 20 cm wide at the heel, or 2 to 7.9 inches. Bigfoot has 15 to 22 centimeters, or 5.9 to 8.7 inches, wide palms and stubby fingers. This cryptid has been described to have unkempt, matted hair that is often seen as either dark brown or reddish. Bigfoot's skin varies between black, brown, and tan; and has a sagittal crest on its head.

Bigfoot has been seen with several facial features: a large brow ridge, a flat, black nose, and thin human-like lips. Its eyes has been claimed to be either brown or red, with its odor being similar to a butch who lives under a bridge. Bigfoot has been claimed to be able to run at speeds of 48 to 56 kilometers per hour, slightly faster than the fastest person in the world, Usain Bolt.

Scientists have tried to debunk this film by using top athletes to mimic Bigfoot's movement, but with no avail. While the scientists have claimed that they were successful, a closer inspection between Bigfoot and the athletes shows a completely different story. It was found out that while the athletes fully extended their legs, Bigfoot did not. Bigfoot also showed that its shoulders are much lower than the athletes, and its arms are disproportionately long. Human beings, it seems, are not suitable to be a suspect for the identity of Bigfoot in the Patterson-Gimlin film. If it is not a human, what was it?

A team of scientists has put together a map of Bigfoot encounters throughout the United States, and a pattern began to emerge: the ecology of Bigfoot is nearly identical to that of bears. Not only that, but it has been discovered that Bigfoot's renowned footprints are extremely similar to the footprints of bears. By experimenting with bears, it has been found that once the bear's hind limb steps over the footprint of its front limb, the bear creates a footprint similar to that of Bigfoot. These two discoveries are currently two of the most used pieces of evidence used to debunk the Bigfoot legend. However, Bigfoot is described as a primate as it resembles apes such as gorillas, chimpanzees, and orangutans. Therefore it would seem far-fetched to take assumptions and state that a species of primate is not a primate at all, but rather a bear. If Bigfoot were actually a primate as hinted by all the evidence given about the creature, then it would create several problems in taxonomy and paleoanthropology. Paleoanthropology is the study of extinct primates, and this is the field required to determine the plausibility of Bigfoot being a primate. This field requires a lot of effort and terms, yet the term "evolution" is still misinterpreted.

SECTION 2: EVOLUTION

To get on the same page, the term *evolution* must be defined, and the best source to learn the definition of the term is in Charles Darwin's *On the Origins of Species*. Charles Darwin is amongst the most famous scientists globally, ranking with Isaac Newton and Albert Einstein. During Darwin's later life, his parents and mentors thought that encouraging his interest in the natural world would be useful supplementary to clerical studies. Then, the study of nature should be a primary interest to a theologian, a person who engages in the study of the nature of God and religious belief, because the Earth was God's primary creation. Therefore, Darwin's parents and mentors arranged for him to join Captain Fitzroy as a naturalist on a five-year voyage around the world from 1831 to 1836 in the ship **HMS Beagle**.

Darwin's journal of his voyage shows his great dedication into collecting numerous samples and describing landscapes, climates and cultures as well as an offhand curiosity that led him to explore the wilderness of uncharted regions, armed only with a geological hammer. When Darwin began his voyage, there had already been an established theory of species adaptation proposed by Jean Baptiste Lamarck, who stated that an individual of a species is able to change and adapt in response to the environment in its lifetime. Darwin's experiences during the voyage convinced him that Lamarck's claim was wrong.

After much reflection and studying the work of other scientists in the span of half a decade, Charles Darwin managed to put the process of natural selection together with the recently developed theory of evolution. Even during the early stages in the late 1830's, Darwin had a strong sense of the controversy his theory would create, and hence he was unwilling to publish his theory. In 1843, after many years of thought, Darwin wrote a two hundred-page summary to be published in the event of his death. Although many friends and colleagues urged him to publish the theory, Darwin refrained until the year 1857. This was the same time when Alfred Wallace presented papers about the same theory to the Linnean Society, where Darwin has also presented his own. The scientists from the Linnean Society have agreed that Darwin had precedence, he thought of it first, and therefore he should get the right to publish his papers first.

In the first edition of On the Origin of Species, published in 1859, the term *evolution* never appears, with the term *evolution* first appearing in the sixth edition. Instead of *evolution*, Charles Darwin's term of choice is *descent with modification*, a simple phrase capturing the essence of evolutionary biology: the study of the transformation of species through time, including both the changes occurring within a species and the origin of new species. For more than a century after the publication of On the Origin of Species, biologists thought that evolution proceeded in a slow period of time. This was a result of Darwin's writing,

"We see nothing of these slow changes in progress, until the hand of time has marked the long lapse of ages" (On the Origin of Species, Chapter Four, 1859).

Many recent studies oppose that however, as they indicate that selection in nature is often strong, resulting in the evolutionary change often occurring very rapidly. Pioneered by the study of Galápagos finches by Peter and Rosemary Grant who documented rapid evolutionary change in the birds from one generation to the next in response to weather-induced environmental changes, the study of real-time evolutionary change in nature now has hundreds, if not thousands, of well-documented examples.

The discovery of the rapidity of evolution has resulted in evolutionary experiments in which researchers can alter environmental conditions and test evolutionary hypotheses over a period of several years. Research in this area involved studies on the colour of guppies in Trinidad. Observing that guppies were generally much more colorful when they lived in streams that lack predators, John Ender moved some guppies from streams with predators to nearby streams that lack predators. Very quickly, the population evolved luxuriant coloration as a result of the females' preference for brighter males. Now without predators, this led to the rapid evolution over 14 generations. Subsequent studies have shown that during this time, the guppies freed from predation evolve may other differences, such as in growth and reproductive rates.

To apply the same to Bigfoot, confirmation on the legendary cryptid being a primate must be confirmed in order to flow with the legends way easier than it naturally would as most, if not all Bigfootologists claim that Bigfoot is indeed a primate. To confirm Bigfoot as a primate, there is an important question to be asked: What is a primate?

GIANT APES: THE CULTURALLY DIVERSE CRYPTIDS

Primates are mammals including lemurs, monkeys, apes, humans and the other related forms. They are characterized by an advanced development of binocular vision, which results in stereoscopic depth perception, specialization of the hands and feet for grasping, and the enlargement of the cerebral hemispheres, better known as the left and right hemispheres of the brain. To adapt to grasping, most primates have five digits in their appendages, including opposable thumbs. Humans are the only known exceptions of having pentadactyl feet that are not prehensile, meaning that the feet of humans possess five toes, but aren't suitable for grasping. The hands of primates are sensitive, adding to their sense of touch.

Unlike many other mammals, primates have particularly flexible and lithe shoulder and hip joints. The shoulders help primates to have overarm movement, ideal for swinging through trees and climbing. Primates' hips are just as mobile, allowing them a greater range of motion in their legs. Besides those mentioned, other primate characteristics include having fingernails and clavicles, also known as collarbones. All primates exhibit the tendency to be erect, being visible when even quadruped primates sit or stand. Most primates also occasionally exhibit bipedalism, the ability to stand on two hind legs. With all that said, Bigfoot shows many of the characteristics of primates.

Just like primates, Bigfoot possesses five digits in its appendages with what seems to be nails instead of claws. As seen in the Patterson-Gimlin film, Bigfoot also has a particularly flexible and lithe shoulder and hip joints relative to an animal its size. Similarly to most primates, Bigfoot has shown to be capable of bipedalism as it is mainly spotted walking on its hind limbs. Based off of photos of the cryptid, Bigfoot most likely has an advanced development of binocular vision, and therefore its stereoscopic depth perception. Based off of all the previously mentioned similarities, Bigfoot is indeed considered as a primate. Finally putting Bigfoot in a group of mammals, we need to further narrow down the placement of Bigfoot based off of detailed description of the cryptid.

Taxonomically speaking, primates include two distinct lineages: strepsirrhines and haplorhines. Strepsirhines are a diverse group of small animas adapted to specialize ecological niches in Africa, Madagascar, and Asia with the term 'Strepsirrhini' meaning 'twist-nose'. The woolly-coated primates possess pointed snouts with moist noses, prominent whiskers, and a divided upper lip tied to the gums by a membrane. The naked most area of skin on the snout on the rhinarium, the snout, is linked to their sense of smell. Mainly nocturnal creatures strepsirrhines have light reflecting discs in their eyes called the tapetum lucidum to aid with night vision. Strepsirrhines are distinguishable from haplorhines because of a number of physiological and morphological features of the inner ear, blood circulation, and digestion. Bigfoot unsurprisingly doesn't belong to the Strepsirrhini family, but to the Haplorhini family.

'Haplorhini' is a term meaning simple nose. Old and new world monkeys, apes, and humans form the suborder Haplorhini. Their facial features distinguish Haplorhines, the structure of their skulls, dentition, and the placenta. Linked to an increasing reliance on vision and a decreasing reliance on smell, haplorhines show a trend towards flattened faces with forward facing eyes as seen in Bigfoot. While strepsirhines have wet noses, haplorhines have dry noses. Haplorhine brains are larger than those of strepsirhines. The lower half of haplorhines' jaws is fused together, and their incisors are spatulate, or shovel-like. With the exception of humans, there is a space between the upper lateral incisor and the upper canine tooth called the diastema. The size of molars increases towards the back of the mouth. Opposing strepsirhine embryos that have many membranes between them and the maternal circulation, haplorhine embryos are in contract with and surrounded by the maternal blood supply.

Knowing that Bigfoot is a species from the Haplorhini subfamily of primates, is it possible to further narrow down the placement of the legendary cryptid? There are three main groups in the Haplorhini subfamily: Tarsiiformes (tarsiers), Platyrrhini (new world monkeys), and Catarrhini (old world monkeys, apes, and humans). Looking at the characteristics of tarsiers, it is not difficult to state that Bigfoot is not a type of tarsier. Opposing what is known about the cryptid, tarsiers are entirely carnivorous, preying on insects, lizards, and snakes. Tarsiers are arboreal, meaning that they live in trees, moving through the forest by launching themselves between the trunks of trees, propelled by the tarsiers' substantially elongated hind limbs. Tarsiers' eyes are more than twice as large as that of Bigfoot relative to their body size. Similarly to the giant sloth theory, tarsiers also have tails, which Bigfoot lacks. The last difference to be discussed is the location. While Bigfoot and Sasquatch live in North America, tarsiers live in southeastern Asia.

Now excluding Tarsiiformes from the list, Bigfoot should either be part of Platyrrhini or Catarrhini. Unsurprisingly, the answer isn't that difficult considering Bigfoot's size and overall shape. In the end, Bigfoot considers as a species of Catarrhini instead of Platyrrhini. Consisting of the New World monkeys, Platyrrhines are small to mid-sized primates. The New World monkeys range from the pygmy marmoset, the world's smallest monkey, at 14 to 16 centimeters in length (5.5 to 6.3 inches) and a weight of 120 to 190 grams (4.2 to 6.7 ounces) to the southern muriqui, at 55 to 70 centimeters in length (22 to 28 inches) and a weight of 12 to 15 kilograms (26 to 33 pounds). Almost all of types of New World monkeys are arboreal with some rarely coming to the ground, therefore the knowledge of them is less comprehensive than that of the more easily observed Old World monkeys. Unlike most of the Old World monkeys, many New world monkeys form monogamous pair bonds, meaning that they have the habit of having one mate at a time, and show substantial paternal care of their young. Some New World monkeys live in very large groups, such as squirrel monkeys, which have groups of up to five hundred members, although these groups may occasionally break into smaller troops. Bigfoot does not show any of these characteristics, and reaches roughly 4.3 times the height of the southern muriqui, taking the factor of the upper estimates in both primates.

The last group of the Haplorhini family to be discussed is the one that Bigfoot is most likely to be in: Catarrhini. Alongside Platyrrhini, Catarrhini is one of the two subdivisions of simians. Simply put, the simians are monkeys, cladistically including apes. Catarrhini, from Ancient Greek meaning "down nose" have nostrils faced downwards as opposed to the sideways facing nostrils of the platyrrhines. Other characteristics of catarrhines are tails unable to grasp, flat fingernails and toenails, a tubular ear bone, and eight premolars as opposed to a dozen. Most species of the Catarrhini group show considerable sexual dimorphism and do not form monogamous pair bonds. Just like the famous Bigfoot, there are some catarrhine species that do not live in social groups. Similarly to the platyrrhines, catarrhines are generally diurnal, have grasping hands, and have grasping feet with the exception of humans.

As mentioned, Catarrhini can be further divided into Old World monkeys and apes. It is a solid fact that Bigfoot is indeed an ape rather than an Old World monkey. The first difference is that while Old World monkeys have tails, apes and Bigfoot don't. Secondly, Old World monkeys live in Asia and Africa, just as referred by their name. On the other hand, there are many apes that have evolved in North America, the overall ecosystem for the primate cryptid. Thirdly, Bigfoot is known to be extremely intelligent, as it is capable of hiding for decades from human contact. The surprisingly high intelligence for an animal could mean that Bigfoot has a very complex brain, a feature that could reach the highly complex brains of apes. As a Bigfoot carcass has yet to be recovered, we are yet to speculate about the other qualities of apes. Nonetheless, the basics of the primate skeleton should apply for Bigfoot.

The primate skeleton can be divided into three parts: the axial skeleton, the forelimbs, and the hind limbs. Predominantly, the axial skeleton is composed of the vertebrae and ribs. On the forelimbs, the main focus is on the clavicle (collarbone), humerus (upper arm bone), radius and ulna (forearm bones), and the phalanges, better known as the finger bones. In a similar concept, the main focus on the hind limbs is on the pelvic girdle (pelvis), femur (thighbone), tibia and fibula (shinbones), and the phalanges. Using these bones, we can guess about how extinct primates were like.

To understand morphological adaptations of extinct primates, paleoanthropologists use a method called the comparative method. As it applies to paleoanthropology, the comparative method is a method of comparing extinct and modern primates based on influences gleaned from the studies of the form and functions of modern primates. If paleoanthropologists were to look at the relationship between certain dental characteristics and diet in modern primates, they'd have to search for identical dental characteristics to a similarly sized extinct primate. When a modern primate is found to exhibit the same dental characteristics as those seen in a similarly sized extinct primate, then the researcher hypothesizes that the two apes must have eaten the same types of food. This process can also be applied to studies of the postcranial skeleton and locomotion in extinct primates.

A fundamental issue in paleoanthropology is the determination of extinct primates' body mass. The body mass is important in paleoanthropology because it relates to many aspects of the biology, behavior, and ecology of primates. In order to determine the body mass of an extinct ape, paleoanthropologists use a statistical procedure known as "correlation", which refers to the relationship between variables. Many features of primates are positively correlated with body mass, such as the cross-sectional area of molars. This indicates that large primates tend to have relatively large teeth. It is hypothesized by paleoanthropologists that positive correlation seen in modern primates also stands true for extinct primate species. Therefore, the discovery of a few teeth from an extinct primate can provide sufficient data to make a rough estimation of its body mass.

Two morphological aspects of dentition are to be discussed, as they are concerned with diet: enamel thickness and dental morphology. Firstly, interspecific variations in enamel thickness tend to reflect differences in the physical properties of food that are eaten by primates. Species with a diet composed of hard, gritty food such as seeds tend to have thick enamel while herbivorous species tend to have thin enamel.

Secondly, although the anterior dentition of many primates serves non-dietary functions such as grooming and defense, it does serve in the initial preparation of food for mastication, also known as chewing. The strongest functional signal for incisors and canines is seen in primates that use their teeth against bark, wood, and fruit. Primates that can open bark and wood to retrieve insects and plant exudates tend to have procumbent incisors, commonly referred to as forward-projecting incisors. Frugivores (animals that specialize in eating fruits) often have relatively larger, broader incisors than the incisors seen in folivores (animals that specialize in eating leaves). The existence of the pattern is because a frugivore first bites into its food using the anterior dentition. This is similar to what most people do when eating an apple or even peeling the rind off an orange. However, anterior dentition also serves social and defensive roles, including threatening open-mouth displays and biting. Due to this, it can be difficult to separate dietary and socioecological roles for anterior dentition in extinct primates. On the other hand, posterior dentition has a stronger dietary component than the anterior dentition. A primate that eats leafy materials passes the food directly into its premolars and molars, meaning that it is easier to use the posterior dentition than the incisors and canines to eat leafy materials. For folivores, the combination of thin enamel and sharp, pointy cusps enables them to efficiently slice leaves, just like a pair of scissors. Insectivorous primates also have sharp cusps on their teeth; yet do not necessarily have thin enamels because they must cut through the hard outer body parts of insects to get to the soft insides. The molars of frugivores tend to have low, rounded cusps called bunodonts, which serve as a broad basin for masticating fruit pulp. Other clues to the diet of extinct primates can be found on their cranial skeleton.

The cranial skeleton is a unique part of the animal as it provides information on both diet and locomotion in extinct primates. The most important dietary aspect of cranial morphology includes attributes associated with chewing. Chewing is achieved predominantly by the actions of the masseter, temporalis, medial pterygoid, and lateral pterygoid, all of which are muscles. Each of the four muscles connects the mandible to numerous parts of the skull. Paleoanthropologists are capable of looking at the size and location of the muscle attachments on the skull and mandible to gain information on the relative bite force involved in mastication.

Primates that feed on hard objects tend to have larger muscles and muscle attachments on their skull and mandible than those that feed on softer objects.

Cranial anatomy and morphology also provide information on the activity patterns and locomotion of extinct primates. Information on the activity patterns can be determined by looking at the relative size of the eye orbits compared to the overall size of the skulls. Primates that have relatively larger eye orbits compared to the size of their skulls tend to be nocturnal, while diurnal primates usually have relatively smaller eye orbits compared to the size of their skulls. Paleoanthropologists also took a look at the location of the foramen magnum, the hole located in the base of the skull through which the spinal cord passes, to approximate body posture and locomotion found in primates. In quadruped primates, the foramen magnum tends to be placed more at the back of the head, indicating a prone body posture. In upright primates however, the foramen magnum is located directly under the skull; which is seen in Bigfoot.

The postcranial skeleton of apes can be examined to deduce the locomotion behavior of extinct primates. The focus will be on the basic postcranial features associated with arboreal quadrupedalism, terrestrial quadrupedalism, leaping, and suspension, the four common locomotion patterns found in primates. Arboreal quadrupeds tend to have a narrow axial skeleton, long tails, moderately long digits, and short forelimbs and hind limbs of equal length; all of which do not fit the description of the famous primate cryptid of North America. Arboreal quadruped primates tend to be in a constantly flexed position, slightly similar to what is seen in the Patterson-Gimlin film. As a result, the elbows and knees of arboreal quadrupeds show specific adaptations to support the animal in a flexed posture. These adaptations and body posture lower the center of gravity of an arboreal quadruped, enhancing stability and balance on an inherently unstable surface.

The limbs of terrestrial quadrupeds are built more for speed than stability. The forelimbs and hind limbs are long, and tend to be held in an extended position. Two other features terrestrial primates tend to have are a short tail and digits. In leaping primates, several postcranial features are related with the strong propulsive forces required by the primates to jump. A main feature found in leaping primates is their long hind limbs, which tend to be much longer than their forelimbs. Leaping primates also have a narrow tibia, which probably indicates the simple hinge-like flexion and extension used by the primates during a leap. On the other hand, suspensory primates have longer forelimbs than hind limbs. Their hips and shoulders are very mobile, ensuring that they can grasp a broad arrangement of constantly changing support structures. Suspensory primates also tend to have very long, curved fingers, which are used to hook onto a branch. Most suspensory primates lack a tail.

It should be noted that the mentioned morphological descriptions represent only a small proportion of the information used by paleoanthropologists to reconstruct the behavior and ecology of extinct primates. Nonetheless, there are several theories that might explain how and why primates have evolved with such features. Out of all the numerous theories given by researchers on why primates evolved in the first place, the most widely accepted surround three slightly different views on primate origins: arboreality, predation, and ecology.

In the year 1972, Frederick S. Szalay originated the arboreal theory of primate origins. In Frederick's theory, primate origins represent an adaptive radiation of new species coming from early arboreal mammals. The momentum for he evolutionary shift was a change in diet from an insectivorous to a herbivorous diet. Grasping hands and feet also evolved in the early mammals in order to ease the movement in the complex web of flexible tree branches. Frederick's theory might account for changes in dentition and limb morphology, but it fails to explain certain aspects of the primate visual system. A feature that remains untouched by Frederick's theory is the convergence and stereoscopic vision exhibited by primates that other arboreal mammals lack.

Because of the untouched feature of primates from Frederick's theory, Mart Cartmill came up with an alternative idea on primate origins in 1992 called the visual predation theory. In Cartmill's theory, primate origins can be traced back to visual adaptations for hunting prey in arboreal habitats. Another suggestion made by Cartmill is that arboreal hunting resulted in the evolution of the primates' grasping hands and nails to capture and hold prey in small, terminal branches instead of claws. Despite explaining a wider suite of characteristics, Cartmill's theory has not been supported by the detailed ecological studies made on insectivorous primates. It has been discovered in 2003 that nocturnal insectivores are more likely to hunt using smell and sound rather than their sight. Furthermore, field studies made by Garber in 1980 revealed that tamarins, which have claw-like nails, are experts at both hunting for insects in terminal branches and clinging to tree trunks to feed on exudates.

The observations made by Garber were part of what led to Robert Sussman's more ecologically based idea on primate origins in 1991 called the angiosperm co-evolution theory. Sussman suggested that the major momentum for primate origins and adaptations was the roughly contemporaneous evolution of angiosperm plants that produced tasty and nutritious fruit in their terminal branches. Field studies of modern mammals similar to those that led to the first primates by Rasmussen in 1990 failed to provide conclusive evidence in either Sussman's or Cartmill's theories. As a matter of fact, field studies provide support for aspects of both theories. Therefore the incomplete fossil record, particularly during the earliest stages of primate evolution, makes it impossible to determine which of the mentioned theories is correct.

Dinosaurs were the dominant vertebrate species for 164.5 million years between 230 and 65.5 million years ago where the fossil record indicates their mass extinction. The reason behind the mass extinction is still heavily debated, although the most prevailing theory is that a colossal asteroid impact threw up a long-lasting global layer of debris, thereby blotting out sunlight and disrupting plant growth. The loss of plant species caused the extinction of large herbivores depending on them for food, and therefore the extinction of large predators. At this point starts one of the most common misconceptions about mammals, which states that mammals evolved after the K-T extinction, which stands for the Cretaceous-Tertiary extinction.

Mammals actually evolved from cynodonts, mammal-like reptiles, roughly 220 million years ago. This means that by the time dinosaurs went extinct 154.5 million years later, mammals were already a long lineage. Thus, the end of the Mesozoic Era, the age of the dinosaurs, marks the start of the Cenozoic Era, the age of the mammals, starting from 65.5 million years ago and still continues today. An era is one of the eleven geologically determined and clearly defined periods of time. Therefore, the geological period is particularly relevant to primate origins and evolutions. The Cenozoic Era is subdivided into seven epochs: Paleocene, Eocene, Oligocene, Miocene, Pliocene, Pleistocene, and Holocene. An epoch is a span of time smaller than an era.

The Paleocene epoch lasted for 9.7 million years between 65.5 and 55.8 million years ago. During that time, the geography was drastically different from the geography today. Africa, India, and the Americas were island continents. Much of Europe and western Asia were flooded, and Australia had just separated from Antarctica. Not only did the geography differ in the Paleocene epoch than what is known today, but also did the climate.

During the Paleocene epoch, the Earth was a considerably warmer and wetter place. Much of the North Hemisphere and parts of the Southern Hemisphere had a warm, temperate climate. Unlike today, Greenland and

Antarctica had cool, temperate temperatures. Just like what is currently being observed, central Africa and parts of southern Asia had hot, tropical climates. Tropical forests covered much of central Africa and southern Asia while North America, Europe, and Eastern Asia. The geography and meteorology during the Paleocene epoch is quite important due to the fact that regions formerly covered in temperate forests now contain some of the best sites for locating Paleocene mammals.

In various parts of the world, paleontologists have found fossilized remains of Paleocene mammals, including rodents and bats. Best represented by a group known as Plesiadapiformes, some remarkably primate-like taxa have been found in North America, Europe, and Asia. Plesiadapids ranged in body mass from 7 grams to 3 kilograms, making the largest plesiadapids over 425 times heavier than the smallest plesiadapids to be discovered. The cranial remains of plesiadapids indicate a nocturnal activity pattern because they have relatively large eye orbits. Studying the teeth of plesiadapids shows that these primates had a primitive dental formula of 3.1.3.3. Based on the studies of their dentition, paleoanthropologists safely claim that most plesiadapids ate seeds and insects. An analysis made on the few postcranial remains that have been discovered suggests an arboreal, scampering mode of locomotion similar to the locomotion seen in squirrels.

Up until about forty years ago, plesiadapids were classified as primates because of certain aspects of their teeth and that their limbs were adapted for an arboreal lifestyle. More recent investigations indicate that plesiadapids are most likely not members of the Primate Order because they retain several primitive mammalian traits, which are the lack of the postorbital bar, claws in their digits instead of nails, eyes placed on the side of the head, and greatly enlarged incisors. Genetic research made by Springer in 2003 however, places the timing of primate origins at around 85 million years, well before the Paleocene epoch. Therefore, either primate remains haven't been discovered from the Paleocene epoch or the interpretations of a few cranial characters in Plesiadapids are flawed. A possible answer to the long-standing issue can be found in a study on primate origins by Bloch in 2007. To be specific, a cladistic analysis of newly discovered and amazingly preserved Paleocene mammals indicates that plesiadapids and a few other contemporaneous mammals are sister taxa to primates of modern aspect, known as the Euprimates.

During the Eocene epoch, which lasted for 21.9 million years between 55.8 and 33.9 million years ago, the tectonic plates continued to move. The movement of the tectonic plates causes changes in the position, size, and shape of the continents and oceans. South America and Africa were still island continents, as was North America during the early stages of the Eocene epoch. The main difference in paleogeography involved land bridges connecting between the major continents.

Early in the Eocene epoch, there was a connection between North America and Europe and between Asia and Europe. Over the course of several millions of years, the land bridges were lost, resulting in the divergence of the various mammals on each continent. The global patterns of continental geography contributed to some of the most extreme climatic conditions of the Cenozoic Era. During the early parts of the Eocene Epoch, the global climate was extremely warm and wet, followed by a slow reduction in temperature and humidity for the remainder of the epoch. As a result, what were once warm, temperate forests have expanded into polar regions. Although many lineages of modern mammals such as ungulates, hoofed mammals, evolved during the Eocene, most species were typically smaller in body size and mass than modern forms. Fossil deposits dated to the Eocene epoch also provide the first evidence of primates similar in form to modern strepsirrhines.

Some extinct mammals recovered from Eocene deposits exhibit morphological features similar to those seen in modern strepsirrhines and even a unique group of haplorhines. The morphological features are a postorbital bar and nails instead of claws. Because of the abrupt morphological transitions from the more primitive features of Plesiadapiformes during the Paleocene epoch, the Eocene primates are often categorized as Euprimates, which means that they look very similar to a few modern primates. Based on several unique characters, most Eocene primates are divided into four taxonomic groups: Adapoidea, Omomyoidea, Eosimiidae, and Oligopithecidae. Adapoidea and Omomyoidea are believed to be the ancestors of lemurs and tarsiers respectively, and are therefore irrelevant to the mystery of Bigfoot. The two groups that are typically believed to be the ancestors of haplorhines, and therefore the ancestors of the bipedal ape cryptid, are Eosimiidae and Oligopithecidae.

Paleoanthropologists have recovered some fragmented jaws, loose teeth, and a few postcranial bones of Eocene primates from China and Egypt. The bones recovered from China are placed in the Eosimiidae family and those recovered from Egypt are placed in the Oligopithecidae family. Eosimiids were generally small animals, weighing up to just a hundred grams, the weight of a newborn fox cub. Similarly to platyrrhines, eosimiids had a dental formula 2.1.3.3. with small, spatulate incisors and broad premolars and molars. The postcranial fossils discovered show properties seen unique in monkeys.

Oligopithecids ranged in weight from about 900 grams to 1.5 kilograms. Surprisingly, oligopithecids tend to have dental formula of 2.1.2.3., the same dental formula found in modern catarrhines. Furthermore, the cranial morphology of oligopithecids such as the shape of their auditory bullae bears some outstanding similarities to modern catarrhines. The diet of oligopithecids might have been composed of insects and leafy materials.

Because these mammals are the first to be definitively identified as members of the Primate Order, resolving phylogenetic relationships in Eocene primates is a fundamental issue in primate evolution. Due to the Eocene primates being seen as the first definitively identified members of the Primate Order, these primates are the most likely to be the common ancestors of all modern primates. Moreover, paleoanthropologists are seeking answers to whether adapoids were the first strepsirrhines and whether omomyoids were the first tarsiers. Part of the problem with resolving issues of primate origins in the Eocene epoch is that some taxonomists only use a few obvious morphological features to define each taxonomic group, such as the size of their eye orbits. This type of scientific debate fails to resolve the phylogenetic relationships given between individual taxa. In contrast, cladistic studies of primates from the Paleocene and Eocene epochs led to the formulation of three hypotheses on primate origins. The first theory is that the research supports adapoids and omomyoids as distinct primate lineages with no connection. Another theory is that adapoids and strepsirrhines form a single group, similarly to omomyoids and haplorhines. Finally, the third theory is that there is indeed evidence that haplorhines have evolved from omomyoids. More recent discoveries of unique adapoids and omomyoids in Asia and Africa could only mean that the three points could only been viewed as hypotheses that will be tested and revised by paleoanthropologists in the future.

Lasting 17.7 million years, the Miocene epoch spanned from 23 to 5.3 million years ago, marking the evolution of primitive apes. The main paleogeographic feature pertinent to the origins of primates involved cycles of expansion and reduction in the size of primates' habitats in Eurasia and the Mediterranean. Land bridges kept forming and disappearing between Africa, Europe, and Eurasia. During the early stages of the Miocene epoch, the climate was similar to today's conditions with the exception of it being slightly warmer than the climate today. About 15 million years ago, the climate became considerably cooler and drier as glaciers formed in Antarctica. As a result, tropical forests transitioned to a mosaic of woodland savannah, made up of trees and grass, and savannas, which are made up of grass. The cooling climate also resulted in the expansion of savannah-woodland environments in Africa and Eurasia. Old World monkeys and apes diverged during this time, and the apes then underwent an adaptive radiation into eighty to a hundred species, roughly four times more than the estimates of the amount of modern ape species.

The early stages of the Miocene epoch lasted for seven million years from 23 to 16 million years ago. During the early stages of the Miocene epoch, ape-like primates evolved in eastern Africa. Out of the many ape-like primates discovered from this time, the most abundant specimens are part of the genus Proconsul in the Proconsulidae family. These diurnal apes ranged in weight from 17 to 50 kg, marking a major increase in body size within the primate clade. Despite the increase in body size, estimates place proconsulids' brain size to be relatively equal to those seen in Miocene monkeys. Proconsulids had a largely frugivorous diet and sexually dimorphic canines. Most proconsulids were quadrupeds with the exception of a few species that were more arboreal than others. Overall, Proconsul exhibits a collection of traits of both apes and monkeys. For example, the traits found in Proconsul that is similar to primates are a relatively large body size, thick molar enamel, and the lack of a tail. Certain morphological aspects of the backbones and pelvis of these primates are just a few of the traits that are similar to those of monkeys.

Lasting 4.4 million years, the middle Miocene epoch lasted from 16 to 11.6 million years ago. During this period of time, African apes moved across land bridges to colonize Eurasia and eventually colonized parts of Eastern Europe and Asia. In 2003, David Begun hypothesized that some European apes migrated back to Africa somewhere in the middle Miocene epoch. As most species appear to have gone extinct roughly 13 million years ago, it is clear that the African immigrants did not last long. Some of the most well-known and interesting apes from the middle Miocene epoch are Dryopithecus from Europe and Sivapithecus from what is now known as India and Pakistan. These primates were big, with their body mass estimated to be anywhere between 20 and 90 kilograms. Similarly to modern apes, Dryopithecus and Sivapithecus had teeth suited for chewing fruit pulp, shortened snouts, and long, strongly built jaws. A shortened snout reflects a reduced reliance on olfaction, a feature found in primates. The cranium of Sivapithecus is found to be similar to that of Orangutan, although their postcrania have little in common. It has been estimated that the brains of Dryopithecus were similar in size and proportions as modern common chimpanzees. Morphological interpretations of the postcrania of Dryopithecus and Sivapithecus reveal that they probably had the same suspensory locomotion seen in modern apes. All in all, both Dryopithecus and Sivapithecus are closer to apes than they are to monkeys.

The late Miocene epoch lasted for 6.3 million years from 11.6 to 5.3 million years ago. During the late Miocene, there was a continuous, gradual decline in global temperatures. In many regions of the world, temperate and tropical forests shrunk in size because of the reduced rainfall and temperatures. Although these changes possibly caused the extinction of many species of apes in Europe and parts of Asia, other species that live in those regions migrated to the tropical zones of Africa and Southeast Asia. As a result, some extinct apes found in middle Miocene such as the previously mentioned Dryopithecus and Sivapithecus survived into the late Miocene. Furthermore, paleoanthropologists have found remains of some newcomers to the ape lineage such as Oreopithecus, Ouranopithecus, Lufengpithecus, and Ankarapithecus. Out of all the Miocene apes, Oreopithecus is arguably the best-known ape because of the abundant fossils found in Italy. Oreopithecus had the typical catarrhine dental formula of 2.1.2.3. with its dental morphology indicating a folivorous diet, a rather remarkable adaptation in the ape lineage. Most modern apes, with the exception for a few species of gorilla, are more frugivorous than they are folivorous. While Oreopithecus may have had a rather large body mass of around 30 kilograms, it had a relatively small brain, a trait that isn't found in apes. Examination of numerous postcranial remains belonging to Oreopithecus indicates a suspensory locomotion similarly to modern apes.

Ouranopithecus was a large ape that lived in what is now known as Greece, weighing in anywhere between 70 and 110 kilograms. Ouranopithecus possibly had a diet consisting of hard, gritty foods. As seen with Sivapithecus, the skull, and therefore the face, of Ouranopithecus is similar to what is seen in modern orangutans. Like most species of primates, the teeth exhibit clear signs of sexual dimorphism with males having much larger canines than those of females. No postcranial bones of this ape have been discovered yet.

The Pliocene epoch lasted for 3.5 million years, starting from 5.3 million years ago and ending 1.8 million years ago. During the Pliocene epoch, a land connection formed between North and South America through the Panama isthmus. The Tethys Sea became isolated and eventually formed part of what is now known as the Mediterranean Sea. Partially caused by the formation of continental land bridges and isolated seas, the climate continued to cool and dry over the course of the Pliocene epoch, further reducing the size and expanse of tropical forests. Temperate regions continued transitioning from forests to grasslands. Because of the geographic and climate changes, the Pliocene epoch marks the end of the "Age of the Apes" and signaled a broad expansion of monkeys as well as the earliest hominids. Pliocene monkeys have been discovered in the Old World, which consists of Africa, Europe, and Asia. Some specimens have even been discovered in southeastern Britain and the Island of Sicily. The Pliocene monkeys are similar in many ways to modern cercopithecines and colobines, two groups that are part of the Cercopithecidae family.

Extinct cercopithecines represent a diverse collection of primates found in Africa and Asia. Out of the ten genera of cercopithecines from the Pliocene that have been yet to be described, five still exist. The genera that still exist are the Macaca, Papio, Cercocebus, Theropithecus, and Cercopithecus. This indicates the success of the radiation that might last for millions of years to come. Body mass estimates of the cercopithecines range from 9.5 kilograms up to 96 kilograms. This group of primates is very diverse, but they generally share several dental, cranial, and postcranial features similar to the features found in modern guenons and macaques, which consist of mangabeys, baboons, and geladas.

Cercopithecines were not the only extinct Pliocene monkeys to be found in the Old World. Recovered from the same regions as the cercopithecines are extinct colobines. Paleoanthropologists have described roughly a dozen genera of extinct colobines from the Pliocene epoch, ranging in body mass from 4 to 35 kilograms. Although some extinct and modern colobines share morphological similarities, many extinct colobines are visually drastically different from the modern members of the subfamily. To be exact, many extinct colobines retain primitive Cercopithecidae characteristics, which are low molar cusps, compared to the high molar cusps used for shearing leafy materials in modern colobines, and postcranial evidence for terrestrially. Modern colobines on the other hand, are nearly exclusively arboreal.

If Bigfoot was a primate, it likely evolved somewhere in the early Miocene epoch, eventually reaching North America. Taxonomically speaking, Bigfoot is most likely a close relative of orangutans as the closest known primate to the cryptid is Gigantopithecus Blacki. Gigantopithecus is understandably the core of one of the greatest theories about the cryptid. However, that might not be the case with the North American Bigfoot due to the overall nature of the two primates. In 1935, paleontologist Gustav Heinrich Ralph von Koenigswald visited a Chinese apothecary shop in Hong Kong and discovered an unusually large molar. Fossils like the large molar are often found in Traditional Chinese medicine where they are described as dragon bones, but studies revealed that the large molar discovered by Gustav came from a type of colossal ape. When Gustav was describing it as a new genus, he named the ape Gigantopithecus that literally translates as "giant ape", a clear reference to the massive size of Gigantopithecus. Ever since Gustav's discovery, over 1,300 teeth have been found, with many coming from the Traditional Chinese medicine market. More exciting are the discoveries of several lower jaws that allowed paleoanthropologists to slightly infer what Gigantopithecus would've been like. Unfortunately, the lower jaws mark as a dead-end as no other parts of the skeleton have been found so far.

Gigantopithecus blacki, the largest species of the genus, is roughly three meters tall and weighs up to 540 kilograms, only slightly heavier than the upper estimates of Bigfoot's size as discussed in Section 1. Both Gigantopithecus and Bigfoot are solitary animals, and are often believed to have reddish brown hair covering their body, further persuading people to believe that Bigfoot is a genus of Gigantopithecus. These two giant primates are mainly herbivorous, and both of their arms are equally disproportionate when compared to the arm of humans. While these all add up to a good theory, it is more reasonable to claim that Bigfoot and Gigantopithecus are very close relatives, but not part of the same genus.

A major difference between Gigantopithecus and Bigfoot is locomotion. Gigantopithecus was primarily quadruped while Bigfoot is mainly bipedal. Another difference is geography. These two apes are located on entirely different continents, with Bigfoot being located in North America and Gigantopithecus being located in China, India, and Vietnam of Asia. Thirdly, Bigfoot is observed to eat more meat than Gigantopithecus, indicating that Bigfoot is likely an omnivore just like common chimpanzees.

Gigantopithecus belongs to Ponginae of the Hominid family. Pongids, the term used to describe the members of Ponginae, are known to be paraphyletic, meaning that they descended from a common evolutionary ancestor or ancestral group but do not include all the descendant groups. Roughly 7 million years ago, pongids gave rise to hominins and hence why they are known to be paraphyletic. Members of this group are sometimes called the great apes, with the most famous example being the orangutan. Although Ponginae has a historical significance, the taxon is rarely used today because there are half a dozen modern species that haven't went extinct. Unlike hominids, which have an upright posture, pongids have a bent over posture, with knuckle walking being common in pongids. Pongids usually have arms that are longer than their legs, which indicate at the adaptation of swinging by the arms. The feet of pongids have low arches and opposable big toes, making the feet capable of grasping. This is unlike hominids that have high arches and big toes that align with the other toes, making them adapted for walking. Pongids are shown to have prominent teeth and large gaps between the canines and the nearby teeth, the exact opposite to what is seen in hominids. Skulls of pongids show that they are bent forward from spinal column, very rugged, and contain prominent brow ridges. Jaws of pongids jut out and are really heavy, the nasal opening is really wide and the facial features are generally sloping. Lastly, there is the brain size. Modern pongids have a brain size of 280 to 705 cubic centimeters, roughly half the size of the brains of hominids, which range between 400 and 2,000 cubic centimeters.

There are major flaws in Bigfoot being a member of Ponginae that must be discussed. First, Bigfoot is more adapted to bipedal walking than pongids as seen by several of its postcranial features. Bigfoot was rarely seen knuckle-walking if at all, and its supposed footprints are more similar to those of hominids. Second is the cranial feature of Bigfoot. Looking closely at the features of its face in the Patterson-Gimlin film, it looked more similar to the description of hominins than pongids. Last, but definitely not least, is brain size. Bigfoot is shown to be highly intellectual, matching and perhaps even exceeding the intelligence of most species of hominins.

This cryptid was described to recognize and escape humans and cameras, capable of staying hidden from us as human beings, and even use advanced tools and ways of communication that could only be seen in hominins. Thus, closest relative for Bigfoot would be in the Homininae family. The closest relatives of Bigfoot would thus be gorillas.

There are four types of gorillas today: cross-river gorillas, mountain gorillas, western lowland gorillas, and eastern lowland gorillas. Although some sources indicate that gorillas started to inhabit Africa 11 millions years ago, it is believed that gorillas separated from other hominids seven million years ago. Eastern and western gorillas, with both having two subspecies, diverged about two million years ago with the Congo River separating the two.

There are several reasons for the division of the subspecies. For example, mountain gorillas separated from the eastern lowland gorilla about 400,000 years ago. According to the hypothesis of experts John F. Oates and Esteban E. Sarmiento, the Cross River gorilla became a new subspecies during the Pleistocene epoch 2.6 million to 11,700 years ago because of adaptation to the scarcity of food during that time.

This would not apply to Bigfoot, however, as it lives in North America. In between 20 and 5 million years ago, Eurasia and North America drifted away from each other. By five million years ago, both continents were nearly in the same position as they are today. Therefore, it would only seem to be logical that the continents have already separated by the time that gorillas evolved. Another feature that would be unsuitable for the theory is size.

The largest species of gorilla is the eastern lowland gorilla, which could reach sizes of 201.2 centimeters, or 6'7", and weigh 249.5 kilograms, or 550 pounds. Taking in the time of 400,000 years since it has been separated from mountain gorillas, then the eastern lowland gorilla would merely be able to increase its size by 0.11-fold. Applying this to its size would give a height of 223.3 centimeters, or 7'4" and a mass of 306.9 kilograms, or 676.6 pounds. This is merely 67% the size of Bigfoot.

Size and location aren't the only problem, however. Both the cryptid and the eastern lowland gorilla have drastically different proportions. For example, the gorilla's arm-to-leg ratio is 31:20 whereas the arm-to-leg ratio of Bigfoot is 37:50. This indicates that while the arms of the gorilla are 1.55 times the size of the legs, the arms of Bigfoot are 1.35 times smaller than the legs. Therefore, the short evolution of gorillas, location, size, and proportions all indicate that Bigfoot is closer to orangutans than gorillas, which would cause some problems as it would be seen in the next section.

SECTION 3: THEORIES

When it comes to misidentifying Bigfoot, bears are nearly always in the top of the list. In the Patterson-Gimlin film however, we face a bit of a problem. Bears are known to be able to walk in two legs, where they are capable of reaching Bigfoot's height. However, bears' forelimbs are not disproportionately long arms compared to humans when standing in two legs. If anything else, the humans would be the ones with the disproportionately long arms compared to bears. As seen in the Patterson-Gimlin film, Bigfoot seemed to walk much more smoothly than bears when seen walking in just its hind limbs. The last evidence opposing the fact that the Bigfoot seen in the Patterson-Gimlin film is a bear comes from the face: bears have a long snout and a distinct, black nose. Bigfoot on the other hand, does not show these features. Instead, it shows a hominid-like face found in primates.

In the November of 2012, former Texas veterinarian Melba S. Ketchum claimed to have proved that Bigfoot is partially human by a DNA sample. Ketchum even went so far as to insist that the government recognizes the cryptids as "*indigenous people and immediately protect their human and Constitutional rights.*" After five years of research about the famous cryptid's claimed DNA by a noteworthy Bigfootologist, Bigfoot has been found to be partially human; who is likely to be from an American origin. A statement released by Ketchum, the lead researcher of the study, reads:

"*Genetically, the Sasquatch are a human hybrid with unambiguously modern human maternal ancestry. Researchers' extensive DNA sequencing suggests that the legendary Sasquatch is a human relative that arose approximately 15,000 years ago.*"

For the study, Ketchum sequenced twenty entire mitochondrial genomes and made use of "next generation" sequencing to obtain three complete nuclear genomes from the claimed Bigfoot samples. As Ketchum's team explained their findings, Sasquatch is a human hybrid with mitochondrial DNA identical to that of humans, while the nuclear DNA contains a non-human sequence. The mitochondrial DNA is the gene inherited from the mother. On the other hand, the nuclear DNA mixes genetic material from both parents. This means that overall, Ketchum's study points to the theory that Sasquatch's origin from crossbreeding a human female and an unknown male species.

It is important to note that the study has not yet been peer reviewed and Ketchum has therefore refused to release her data, explain her methodology, or clearly state where she has obtained the supposed DNA samples of Sasquatch to begin with. According to Houston *Chronicle* science writer Eric Borger, Ketchum has credibility issues from her company, DNA diagnostics, has received more than two dozen customer complaints, and has been given an F from the Better Business Bureau. As stated by a blogger and Bigfoot enthusiast called Robert Lindsay, earlier drafts of Ketchum's study claimed that the DNA of the male of the unknown species came from angels. Unsurprisingly, the scientific community remains doubtful. At *NeuroLogica Blog**, Yale neurologist Steven Novella wrote the following:

"*The bottom line is this: Human DNA plus some anomalies or unknowns does not equal an impossible human-ape hybrid. It equals humans DNA plus some anomalies.*"

This theory has yet to be either confirmed or debunked as the evidence remains hidden from the public. A theory that is as perplexing as that of Melba Ketchum is that Bigfoot isn't a primate, but a giant ground sloth.

A giant ground sloth discovered in North America is Glossotherium. Roughly 3 million years ago, Glossotherium traveled from South America over the newly reformed land bridge to where it was discovered in the Rancho La Brea tar pits, Los Angeles. The tar pits have yielded and preserved outstanding specimens of the giant sloth; which makes it easier to study the creatures.

Glossotherium was bulky, possessing a large head and a heavy tail. Similarly to its relatives, Glossotherium's long, clawed feet were turned inwards. This causes the giant sloth to walk on its knuckles as seen in modern gorillas. Judging by the plant remains preserved in its fossil droppings; Glossotherium seems to have lived on desert shrubs. The giant sloth was capable of rearing up on its hind limbs and used the long claws of its hind limbs to grab food and pull the food towards its mouth.

While Glossotherium could reach to a height of two meters when standing in its hind legs, there are still a lot of flaws in this theory. The first and most important flaw is the anatomy. Glossotherium possessed a tail one quarter of its length, which something Bigfoot lacked. Bigfoot also seemed to have a flat face with features similar to those of chimpanzees. In the other hand, Glossotherium seemed to have a long snout and possessed features far from matching to those of Bigfoot. Another difference is that Bigfoot's feet and hands are more similar to those of primates than those of giant sloths. Giant sloths such as Glossotherium have been proven to be primarily herbivores while Bigfoot has shown to be omnivorous.

Other than anatomy, a major flaw relating to the first would be the geological timeline. Supporters of Bigfoot and this theory would state that Glossotherium would've evolved into Bigfoot. However, it is both geologically and biologically impossible. Bigfoot has shown to be drastically different from the sloth in many ways; with the main difference being moving stance. Glossotherium walked on all four legs, and used its tail for balance, especially when rearing up on its hind limbs to eat. Bigfoot however, has completely lost its tail and is shown to walk on its hind limbs much more than the majority of primates. This feature took millions of years to adapt, while the giant sloth went extinct 11,000 years ago. It would only seem to fit if Bigfoot would be some sort of ape.

The fourth theory to be discussed is a truly bizarre theory. Numerous people believe that Bigfoot is actually a caveman after a witness gave extreme details to his encounter with Bigfoot. In 2007, a resident of Vancouver Island by the name of Robert Wilson claimed to have witnessed the legendary ape. He described it as:

"...What I thought was bear. I drove down and saw what I can only describe as a large, hairy man who looked cave man-like... with sort of Neanderthal features. As big as a bear, easily."

Expanding on Wilson's claim, a 2011 History Channel documentary proposed, *"[the] Sasquatch might not be a giant ape at all, but could be a species of prehistoric human."* This theory might be a reasonable theory, but it does have two major flaws that debunk the entire theory: the migration and biology of Neanderthals.

Neanderthals have existed for 350,000 to 300,000 years ago, long before the arrival of humans in North America. Genetic analysis has indicated that early humans began to migrate across the ancient Beringia, the region that once connected Asia and Alaska, about 25,000 years ago. The earliest evidence of human habitation however, comes only 15,000 years ago.

Neanderthals have only been discovered in Europe and Asia, with no evidence hinting at its existence in North America. This means that the geological timeline and fossil evidence support the fact that Neanderthals could not exist in North America. However, does the biology of Neanderthals match that of Bigfoot?

The simple answer is no, due to many factors. Bigfoot has been estimated to be two to three meters tall, or roughly 6'7" to 9'10", while Neanderthals are estimated to be 1.66 meters tall or 5'5". Another difference between the two is habitat. Unlike Neanderthals who lived in grasslands and woodlands, Bigfoot is restricted to forests and woodlands due to their rather bashful nature. Thirdly, their diets heavily vary. According to video footage and sightings of Bigfoot, it seems to be omnivorous, although Bigfoot is more of a herbivore than a carnivore; only eating meat in special occasions. It is known to Bigfootologists that fruits seem to be the gentle giant's favorite food, especially apples. Neanderthals in the other hand, are mainly carnivorous. One of the unique features of Neanderthals is their weapons, which they have mainly used to aid in their hunts; but that is not the only feature that makes Neanderthals successful hunters. Neanderthals were stocker and far stronger than modern humans, which heavily aided them. Their dense build helped them cope with the cold environment, and their considerable strength was required to hunt large animals such as the Woolly Mammoth.

One last piece of evidence that debunks that Bigfoot is a Neanderthal is the hair. While Neanderthals were hairy, their hair didn't cover their entire bodies like Bigfoot. Instead, Neanderthals seemed to have more or less hair than the well-known Norsemen Vikings. Arguably the most famous of all these Vikings is Leif Erikson, who he himself has described the Bigfoot to be extremely hairy. The difference in the amount of hair found on Bigfoot and Neanderthals can be explained by science.

The most important function of hair in mammals is the insulation against the cold by conserving body heat. Differentiation in the colors and color patterns in hair coats can also serve the purposes of camouflage and sexual recognition and attraction among the members of a species. Bigfoot exists in habitats with climates cooler than that of Neanderthals; meaning that it would need a higher number of hair follicles filling its body than Neanderthals. The differentiating climate however, is only one of the two main factors on the reason behind the difference of hair found in Bigfoot and Neanderthals.

The second important factor is an advanced thermoregulatory system found only in mammals: sweat glands. There are two types of sweat glands: eccrine and apocrine sweat glands. The eccrine sweat gland, which is controlled by the nervous system, helps in the regulation of the body temperature. When internal temperature rises, the eccrine glands produce water into the surface skin, removing the heat by evaporation. Eccrine glands are major thermoregulatory devices if they are active over most of the body, such as in horses, bears, and humans. In other animals such as dogs, cats, cattle, and sheep, eccrine glands are only active on the pads of the paws or along the lip margins; and may be entirely absent over the rest of the body. Such animals often depend on panting for effective temperature control. Smaller mammals such as rodents and shrews can't endure dehydration and hence possess no eccrine glands at all.

Apocrine sweat glands, usually associated with hair follicles, continuously produce a fatty sweat into the gland tubule, a tubule that removes specific substances from the blood, alters or concentrates them, and then either allow them to leave for further use or get rid of them. Emotional stress causes the tubule wall to shrink, letting out the fatty secretion to the skin, where local bacteria break it down into odorous fatty acids. In modern human beings, apocrine glands are concentrated in the underarm and genital regions. The apocrine glands are inactive until they are stimulated by hormonal changes in puberty. Apocrine glands are more numerous in other mammals. Certain specialized glands such as mammary glands, wax-secreting glands of the ear canal, and many mammalian scent glands probably developed from modified apocrine glands.

The difference between Neanderthals and Bigfoot is that the eccrine sweat glands of Neanderthals have developed to be much more efficient than Bigfoot; making it easier for Neanderthals to control their body temperature and adapt to the climate. So all in all, Bigfoot is neither a primate, a primitive human species, a crossbreed of a female human and an unknown creature, a bear, nor is it a giant ground sloth. So what is Bigfoot? That answer might come with the most plausible theory about Bigfoot, bear hybrids.

Ever since the first reported shooting in 2006, unusual bears that are a mixture of both polar and grizzly bears have been found in the Canadian Arctic. After the 2006 incident, at least eight further sightings have followed and have actually been confirmed to be hybrids of polar and grizzly bears by DNA testings. The bears have a cream or light tan fur, intermediate claws, a slender bear snout, and the broad, muscular shoulders of a grizzly bear. Andrew Derocher, a polar bear scientist from the University of Alberta, Canada that studies polar bear populations in the Canadian Arctic and Hudson Bay, states the following in an interview with ScienceNordic:

"We've known for a long time that hybrids between polar bears and grizzly bears were possible. We've known from zoo studies in Europe that you can take a male and female from either species and hybridize them, and their offspring are

fully fertile. To date, all the confirmed hybrids are in Canada. But that doesn't mean that they couldn't exist in Russia for example, where these species come very close to each other, or in Alaska, where they also overlap."

This quote indicates that the Russian Almasty, one of the many cryptids belonging under the Bigfoot family, might also be a grolar bear, one of the names of the polar-grizzly hybrids. A question left unanswered by Derocher is the cause of the two species overlapping. While polar bears are found in the North Pole, grizzly bears are mainly found in tundra. As the two species have adapted in two separate biomes and regions, polar and grizzly bears shouldn't be found in the same area. So what caused grizzly and polar bears to interact to start with? The answer is simple: global warming.

Global warming is partially responsible because of the increase of the Earth's global climate, which shrinks the ice in the Arctic ice where polar bears live and hunt in. Consequently, polar bears have emerged in grizzly bears' territories as the polar bears that live and hunt in the Arctic sea ice are forced on land during mating season in spring and summer. Meanwhile, male grizzly bears are expanding their habitats, hence their appearance in polar bear territories and emerging from hibernation earlier in the years. Evidence supporting the fact that grizzly bears began to arrive in the Arctic due to global warming comes from Inuit hunters. Inuit hunters have long seen grizzly bears in the Arctic, ranging to a span of decades, but it is believed that it recently increased, causing males to further diverge in search of a female. Consequently, the interaction and breeding of the two species has become more common after global warming.

There are two key factors to be discussed about the evolution of the grolar bear: natural selection and genetic inheritance. Natural selection, although a key evolutionary mechanism, is biased. In the natural selection mechanism, individuals that possess characteristics best adapted to their surrounding environment have a better chance of survival and pass on the favorable traits to the next generation. This causes the distinct features between different species of the same family, such as the relative lack of hair and superior brain of humans compared to other primates.

Natural genetic variation within the populations in an environment produces several differences such as size and color, and some of the differences might help in promoting survival. For example, the white coloration of polar bears' fur provides a camouflage that helps in hiding from other polar bears and their targeted prey. This leads to the survival to the polar bears with the better camouflage and their reproduction, the adaptations of those polar bears will be passed on.

Author Name: Alwaleed Alghanim

If the environment were to change over time as seen in the drastic change in the habitats of polar bears, a different coloration might be more beneficial, and hence natural selection will ensure another change. In the case of grolar bears, the change comes from interbreeding with grizzly bears. Populations of the same species might split up after a movement of the tectonic plates, called the geological rift. Each of the populations that split up will eventually adapt to the now slightly different conditions. This might to lead one species becoming two in a process known as speciation. Speciation is the same process described by Darwin about the Galápagos finches, also known as Darwin's finches.

Global warming does answer the question of natural selection in grolar bears, but then there is genes and inheritance. Grizzly and polar bears are clearly two different species, and therefore it would logically seem to be very unlikely, if not impossible, to interbreed these two species into a single, healthy cub. That is actually inaccurate as it is less likely to bring a healthy child by means of incest than interbreeding. A study of Czechoslovakian children whose fathers were first-degree relatives showed the consequences of incest when compared to children that haven't come from incest parents. 47% of the children born from incestuous sexual intercourse were born completely healthy. 42% of the children were born with severe birth defects and/or suffered death at an early age. The remaining 11% of the children were found to be mildly mentally impaired. When the same mothers were impregnated by a non-relative, studies revealed that the 42% of children born with severe birth defects was lowered by 35% to a mere 7% chance of getting born with a birth defect. Therefore, it can be concluded by the studies that incestuous intercourse increases the chances of children being born with severe birth defects by 600%. With an increase in the chances of getting born with a rare genetic disease and even death at a young age, many who are aware of the effects of this act consider incest as both harmful to the child and disgusting. All in all, incest is indeed the act in which many people would misinterpret the effects of interbreeding for.

There are actually reports of hybrid animals other than grolar bears that prove to be way much more successful than incest. Five of these hybrids are zebroids, coywolves, wholphins, ligers, and cama. Zebroid is the collective name for any zebra hybrid where a male zebra is crossed with a female animal from the equidae family, simply known as the horse family. Zebroids never occur in nature, and many of them can be born with a type of dwarfism, almost always being infertile. There are many different animals found in the Zebroid group, such as the Zorse (zebra and a horse), Zonkeys (zebra and donkey), and the Zoni (zebra and pony).

Coywolves are a hybrid between a coyote and a wolf, regularly occurring in the wild. It is so common that all recorded red wolves have been discovered to have coyote genes in their lineage. It is unclear whether the inbreeding is caused by human development limiting the natural habitat of the two predators or if the red wolf as always been a hybrid. The hybrid has caused a lot of problems in the taxonomy of Canids as hybrids are not usually referred to as a different species, though an agreement was made that the red wolf is a subspecies of the wolf, leaving its Latin name not referencing its coyote genes.

Wholphins are remarkable hybrids coming from a bottlenose dolphin that had a successful pregnancy from an orca, another type of dolphin. Wholphins have been known to occur in the wild, with only two living wholphins being in captivity in the Sea Life Park, Hawaii. The first wholphin in Sea Life Park was Kekaimalu, who proved to be fertile when she gave birth at a young age. Kekaimalu's first two babies unfortunately did not survive for a long time, but her third baby fortunately did. Within two months of the baby's birth, who is named Kawili Kai, she became as large as a full grown bottlenose dolphin, which can grow between two and four meters long and could weigh anywhere 150 and 650 kilograms. Kawili Kai now happily lives with her wholphin mother and bottlenose father at the Sea Life Park.

The Liger is a hybrid between a male lion and a female tiger; therefore both its parents are from the same genus, Panthera, but different species. Ligers are currently the largest out of all the big cats, with a length nearly as much as the length of a lion and a tiger combined. These large cats carry characteristics from both parents, such as the love of swimming from tigers and a highly social behavior from lions. Ligers currently only live in captivity as lions' and tigers' territories don't overlap, but there are stories of ligers found in the wild in history. These hybrids were long thought to be infertile, only for the theory to be disproven in 1953 when a fifteen-year-old liger successfully mated with a male lion. Despite having poor health, the cub survived to adulthood. In Jungle Island, a theme park found in Miami, a liger called Hercules can be seen. Hercules is a massive liger with its weight exceeding 410 kilograms, and hence holds the Guinness World record for the largest big cat in the world. Hercules is expected to live a long and happy life as he is extremely healthy.

Cama is a hybrid created by breeding a male Dromedary camel with a Lama in a laboratory in Dubai. The Cama was made with the purpose of having an animal with the size and strength of a camel but the easier temperament and higher wool production of the lama. Interestingly, the cama are one of the few hybrids that are constantly fertile and therefore resulting from both the dromedary camel and the lama containing the exact same amount of chromosomes. As the camel is 6 times larger than a lama, the only method to obtain a cama is by artificial insemination, with only six cama births being successful.

As indicated by these examples, there are two types of hybrids, which will be called the natural and artificial hybrids. Natural hybrids, such as wholphins and red wolves, are the hybrids that occur in the wild without any human interference. Artificial hybrids on the other hand, are hybrids that are bred by humans and often require artificial insemination. Examples of artificial hybrids are cama and all members of the Zebroid family. Grolar bear can therefore be referred as a natural hybrid.

There are two main differences between natural and artificial hybrids: taxonomy and geography. Most, if not all, natural hybrids are composed of two species of the same genus. For example, ligers are made up of tigers and lions, which are two separate species of the genus Panthera. The red wolf is another example as both wolves and coyote are two different species of the genus Canid. While some artificial hybrids such as Zebroids are composed of two species of the same genus, they occur because of cloning and artificial genetic mutation instead of natural selection as seen in natural hybrids. Second, the geography of the two species composing a single natural hybrid must overlap. The best example is once again the liger. Ligers used to occur naturally only because the territories of tigers and lions used to overlap, hence causing the two species to interact. After human development however, the habitats of the two species began to shrink to the point where their territories no longer overlap. As discussed, the territories of grizzly and polar bears have overlapped, hence further supporting that the grolar bear is a natural hybrid.

The capability of interspecies intercourse of the same genus lies within genes and inheritance. Particular traits of the parents pass on to their offspring via the transmission of genetic material. Genes preserve all the information required for the replication of a cell's structure and its maintenance, which is encoded within the deoxyribonucleic acids (DNA). Therefore, genes are the basic units of heredity. Individual chromosomes hold thousands of genes on long strands of DNA. During sexual intercourse, sperm and egg cells fuse to produce two complete sets of gene-bearing chromosomes, with one copy from the father and other from the mother.

Species of the same genus have similar enough DNA to allow this process to successfully occur. The most famous example comes from chimps and humans. Studies reveal that the DNA of chimps and humans are similar by 98%. Other studies show that species of the same genus have anywhere between 70% and over 95% of similar DNA; hence this can be applied to grizzly and polar bears.

Natural selection and genetic inheritance might both support the theory that Bigfoot is no more than the grolar bear; there are two major flaws yet to be covered about this theory, the Patterson-Gimlin film and intelligence. As previously mentioned, the biology of Bigfoot is seemingly different from that of a human; therefore the locomotion of Bigfoot is slightly different from that of a human. However, there is evidence to oppose the fact that Bigfoot is a primate. In 2002, Philip Morris claims that as the owner of Morris Costumes, he had made a gorilla costume that was used in the film. As those that believe in Bigfoot's existence are skeptical because of Bigfoot's movement, Morris responded with the following statement,

"The Bigfoot researchers say that no human can walk that way in the film. Oh, yes they can! When you're wearing long clown's feet, you can't place the ball of your foot down first. You have to put your foot down flat. Otherwise, you'll stumble. Another thing, when you put on the gorilla head, you can only turn your head maybe a quarter of the way. And to look behind you, you've got to turn your head and your shoulders and your hips. Plus, the shoulder pads in the suit are in the way of the jaw. That's why the Bigfoot turns and looks the way he does in the film. He has to twist his entire upper body."

The movement was clearly proven to be a hoax by Morris' statement, as did the overall anatomy. Indirectly speaking, Morris has indicated that the shoulder pads and head of the gorilla suit change the appearance of the person who is supposedly behind the suit. However, the two filmmakers contradict that as Robert Gimlin has stated,

"It ruined me. They'd [the public] come driving in my driveway all times of the night and go 'Bob! We want to go out Bigfoot hunting!" My wife was a teller at savings and loan institution. Of course, she was sitting right there and the public would come in and make smart remarks. They went on and on and on until she comes home crying. She'd say 'I'm not tough enough.' A couple of times we were going to split up over this. I can understand why they don't believe in it because I didn't believe it either, but I saw one, and I know what I saw; and I know it wasn't a man in a suit. It couldn't have been."

Greg Long, the author of The Making of Bigfoot, stated to *Outside Online* one of the most seething criticisms about the Patterson-Gimlin film, stating the following,

> "I'm going to be blunt with you, I consider Bob Gimlin a liar. I think he's a con artist... They [Patterson and Gimlin] need it to be real, [they] are driven emotionally, I believe, to find Bigfoot."

There are two persuasive costume descriptions that support Greg Long's claims. Bob Hieronimous subsequently agrees that the costume used in the Patterson-Gimlin film was just a simple gorilla suit manufactured by Morris Costumes in North Carolina in North Carolina during the 1960's. In the second version of Bob Hieronimous' confession, the fur of the creature was dynel nylon threads stitched to a woven cloth backing, which is nothing like the hide of a dead red horse as previously believed by Hieronimous. The previously mentioned Philip Morris prompted the dramatic modification of Hieronimous' original confession. Morris had also wanted to take some credit for Patterson's footage somehow, and possibly market a new costume. Although Philip Morris has no records of selling a costume to Roger Patterson, the filmmaker who died of cancer in 1972, Morris is nevertheless certain that he sold the costume used in the film. Morris claims that a man named Roger Patterson had bought a gorilla suit that had been on a discount. For some reason, Morris remembers the sale, yet there is no record of it. Philip Morris even claims he had recognized the gorilla costume when he saw the Patterson-Gimlin film on TV.

After an unspecified time from Hieronimous' second version of the costume description, he and Morris came up with a new costume replica of Bigfoot in the Patterson-Gimlin film. Morris has stated that he is able to make more duplicates of the costume for seven to eight thousand dollars. This costume is described as a custom-made, latex-enhanced, and non-stretching eccentric creation. Morris' costume is drastically different from both the Hollywood special effects that are said to have been required, as well as the simple gorilla costume that Philip Morris had originally claimed to be sold to Roger Patterson in the 1960's. This new costume matches Bigfoot seen in the film and could explain the anatomical features and locomotion of Bigfoot as it too has feet made in a similar manner as "long clown's feet" as previously described by Philip Morris. Overall, whether the creature in the Patterson-Gimlin is genuine or a hoax remains unknown. The evidence however, leans towards the fact that the film is a hoax.

Now debunking the first major flaw seen in the grolar bear theory, there is still the second flaw that can be only answered using science, intelligence. As observed by many 'Bigfootologists', whom of which should be accurately named 'Bigfoot cryptozoologists', Bigfoot is capable of using tools for hunting and communication similarly to modern primates. This statement can also be debunked by the study of tool usage and communication in apes. While bears are indeed intelligent creatures, assuming that they are equal to apes in intelligence is absurd, as the following evidence would suggest. Many famous examples of tool usage in primates come in one of the closest relatives to modern human beings, the common chimpanzee.

In the early 1960's when a former secretary named Dr. Jane Goodall was alone in the jungle watching wild chimpanzees, she reported observations of chimps altering sticks and twigs by stripping off the leaves, then using the tools to poke into termite mounds. The termites in the mounds would attack the intruding stick by grabbing onto it with their pincers. With what seemed like an effortless swish, the chimp would pull the stick out of the mound and slide it across its mouth. Following Goodall's report of the chimpanzee's tool usage at her study site located in Gombe Stream Reserve in Tanzania, other sites in various African countries started to report the usage of different types of tools. Across the different toolkits, the various types of tools used, and tool usage documented amongst chimpanzee communities that have been studied for a long period of time, over 35 types of tools have been identified so far. An example of tool usage found in chimpanzees is pestle pounding. In Bossou located in Guinea, West Africa, chimps use this clever technique to process the center of the crown of an oil palm tree in order to eat the pith. Chimpanzees would start by using their hands and feet to prize apart the uppermost branches of an oil palm tree. They would detach a palm frond to use it as a pestle in order to pound and soften the pith at the center of the crown. Finally, the chimpanzee would scoop the pounded pith up.

More recently in several different locations in Africa, chimpanzees have been observed to use stone tools such as hammers and anvils in order to crack open oil palm nuts. Oil palm nuts contain an outer shell so hard that chimps would not be able to access the rich food source without the use of stone tools. However, not all chimpanzees are observed o use stone tools, even with the abundance of the oil palm nuts in their habitats. Comparisons of all chimpanzee sites across Central and Western Africa as well as the respective toolkits of the chimpanzees revealed that chimpanzees may have rudimentary cultures since chimpanzees use a range of unique tools in each site that aren't found in other sites. Therefore, it is suggested that chimpanzee groups acquire expertise with the kinds of tools within their specific group by observing their mothers.

Detailed observations made on wild chimpanzees show that there are differences in tool usage between males and females. Female chimps are yet the most consistent toolmakers and users. Therefore, females depend on a diverse amount of tools made and used creatively in order to obtain a significant portion of their food. Adult males on the other hand, do not seem to use tools that often, hence their independence on tools to obtain food. The young, whether they were male or female, observe their mothers using tools over the course of three to seven years and practice on copying her techniques. In their early attempts, the young would often by unsuccessful, only becoming proficient years later as adults.

Nonetheless, it came as a surprise that the pygmy chimpanzee, the closest living relative to humans, has not been observed to use tools in its natural habitat. This counts as a major difference between the pygmy chimpanzee and the common chimpanzee, as common chimps are some of the most efficient species in tool usage when excluding humans. Some clues from the habitat of the pygmy chimpanzee, also known as the bonobo, might help explain why tool usage is not found in the wild. First, bonobos are more arboreal than the common chimps, meaning that they live in trees. Therefore, bonobos might find all the food required for survival in among the trees and bushes in their territory with the use of tools being unnecessary. Second, bonobos have not been observed eating meat, meaning that bonobos have a diet consisting of fruits, eggs, and foliage. This diet is quite sufficient even without the added protein sources found in common chimps such as termites, ants, and small mammals.

Object manipulation and tool usage in orangutans was shown to be a prominent skill among those that are kept in captivity and similar skills soon appeared in orphaned orangutans that were brought to rehabilitation centers in Borneo. However, these observations revealed a prominent puzzle because the long-term study of wild orangutans had not been observed to use any type of tools. Wild orangutans are fairly solitary, which is necessary for survival in order to avoid competition for food with other orangutans since food resources were scarce and widely distributed. More recent observations of wild orangutans on the island of Sumatra prove that orangutans do indeed use tools. While there are differences in the orangutan populations found in the islands of Sumatra and Borneo, the overall physical attributes of the two populations are very similar. Because of this, it has been believed that the orangutans of Sumatra behave just like those of Borneo. This was not the case as field researchers who withstood two-hour hikes to and from the orangutan site, going waist-high in water, and surrounded by leeches and flocks of mosquitoes, discovered startling differences between the Sumatran and Bornean orangutans.

On Sumatra, orangutans have been observed to use several types of tools for extractive foraging in a highly productive swamp habitat. In order to get into the protein-rich seeds and pulp of the Neesia fruit, Sumatran orangutans are required to use stick tools. Without stick tools to break into the fruit, orangutans aren't capable of breaking into the fruit and get the pulp that is rich in both fats and protein. Using hands or teeth to break into the fruit would be very painful for the orangutan since tiny needle-like hairs surround the edible seeds. Consequently, tool usage allows the orangutans to exploit a valuable food resource that is not accessible to other animals. Orangutans have also been observed using different types of tools than sticks. For example, orangutans have been observed to use a specific type of sticks when opening fruits but they modify different types of sticks in an efficient manner in order to reach honey or insects deep inside tree holes. Unlike in other environments where tool usage provides a significant contribution to the animals' daily calorie intake and therefore necessary for their survival, swamp forest habitats in which Sumatran orangutans live in are rich. However, tool usage allows for greater variety and access to other nutritional food sources.

The scientists that discovered the tool usage in orangutans wondered if orangutans used tools as long as there were Neesia trees, hence they made observations of orangutans living across the river from their original site in a smaller swamp. Neesia trees were found in the smaller swamp, but there was no tool usage, hence the fruits not being part of their diet as opposed to those in the original site. This difference indicated that tool usage among the orangutans of the bigger swamp was a cultural tradition that is observed by each new generation of orangutans, incorporating into their behavior. The orangutans on the other side were unable to get through the river, resulting in the lack of opportunity to learn about tool creation and usage from others in their own community. Another striking difference is the extent of the Sumatran orangutans' social interaction. Sumatran orangutans are highly sociable animals, gathering in groups containing up to 100 members. Before the discovery was made in Sumatra, such a large gathering in between of orangutans was unheard of. In Sumatra, orangutans of all ages foraged together, sharing food and using tools when necessary. None of these behaviors were documented among the Bornean orangutan populations.

In 2005, field researchers reported the first observations made of tool usage by gorillas. The team had been observing Western Lowland gorillas in the Congo for over a decade, with the gorillas never being seen any type of tool usage with the exception of two incidents. A researcher from the team was shocked to see an adult female nicknamed "Leah" wade into a deep water holes left by elephants. When the water reached up to Leah's chest, she reached above her for a branch and broke it off, using the stick as a walking stick. Leah was seen to use the long branch to test of the depth of the water, monitoring the depth of the water until she had walked a full 33 meters (108 feet) before she returned to the shore. The second observation was when Efi, an adult female from another group, was seen taking the stump from a bush and using it as a seat while she used a long stick to dig for herbs.

Each community from most individual species of apes has its own unique set of tools. There are a few problems when it comes to Bigfoot that is clear when looking at the previous examples. First, there is more than one community of Bigfoot. As observed by the orangutan example, this would cause a wide diversity of tools used by Bigfoot with some communities not even capable of using tools, which is completely the opposite of what is been observed by the so-called "Bigfootologists". Secondly, there should be differences between tool usages between the two genders as observed by the chimpanzees. Assuming that Bigfoot has the same male-to-female ratio as humans, then it would have a ratio of 1.014:1. With over 3,300 reports globally, there is less than a 0.03% of all reports being of a single gender. While this highly supports the fact that both male and female Bigfoot have been observed, all of the encounters of tool usage reported about Bigfoot have a consistent toolkit, defying both the chimpanzee and orangutan examples. Thirdly, all of the examples show tool usage found in highly sociable apes. Bigfoot on the other hand, is mostly solitary. Therefore, Bigfoot can be connected to the orangutans of Borneo due to the solitary life in the wild. Unlike what has been observed in captivity, Bornean orangutans have not been observed to use tools in the wild. With all of this evidence against tool usage found in Bigfoot, it is safe to say that Bigfoot is not capable of using tools.

Other than tool usage, the communication of primates must also be studied. The first primate to be discussed is none other than the chimpanzee. Chimpanzees are extremely social animals, living in communities within a specific home territory defended by the adult males of the group. The social structure of chimpanzees is difficult to characterize, but the basic unit of a chimpanzee community is a mother, her infant, and other young offspring. Chimpanzees live in fission-fusion societies, meaning that animals are entering and leaving the entire time. However, chimps identify with a specific community, and always return to their group after their forays away. Whenever one or more chimpanzees return to the community, the community becomes very excited, hugging and kissing those that have returned while shrieks of joy and excitement fill the air. Chimpanzees are renowned to be highly intelligent animals, making it possible that chimps evolved remarkable capacities to live within dynamic and challenging communities that require the recognition of every member, recall the members that have left with some incidents spanning a considerable span of time, comprehension of the complexities of the social hierarchy and rule structure of their community, and possessing the necessary attention and observational skills to learn the numerous types of tools documented within wild chimpanzees.

Leadership of a community is often comprised of related adult males such as brother, and perhaps another adult male with whom the other leaders have a social bond. The dominant males keep the peace in the community and are constantly on the watch for predators such as leopards or human intruders, their biggest enemy. Averaging on every four days, males make a complete border patrol around the community's territory for possible attacks of incursions of males from neighboring communities that might be seeking females. Chimpanzee mothers do not travel as often as males, tending to forage within the central portions of their territory. Unlike many other species of primate, young, sexually mature female chimpanzees leave their community and move to another community to ensure that incest will be avoided.

Chimpanzees possess a remarkable collection of communication strategies, including naturally occurring gestures that are highly similar to those of humans in some cases, facial expressions, body postures, and a collection of vocalizations. Field researchers have found that chimpanzees do not simply produce vocalizations that are related to their emotional state, but are rather to provide a more considerable amount of information through their calls than previously thought. As a matter of fact, the chimpanzees' ability to recognize the meaning and identity of vocalizations that might come from relatives or strangers might be a matter of life and death. If a lone male meets a border patrol from another community, usually consisting of five to seven adult males, the lone male might not be able to escape. Field researchers have come upon single dead chimps with their bodies often showing evidence of cooperative killings by other chimpanzees. Thus, young chimpanzees must learn the members within their group, the nature of what particular vocalizations represent and the source of the call as part of their socialization. All of this requires an immense brain, and similarly to humans, chimpanzees have the capacity to learn much more than what is required to survive in the wild. This behavioral flexibility and learning capacity sets the great apes and humans apart from other primates.

As scientists investigated deeper into the full potential for chimpanzees' vocalization to include more than just the chimpanzees' emotional state, more recent findings appeared from field researchers conducting studies in several East African sites including the Kibale National Forest, Uganda and the Mahale Mountains National Park, Tanzania. Some of the chimpanzee's vocal ranges include screams, hoots, loud and soft alarm calls, and shrieks. Chimpanzees' vocal ranges are estimated to include at least fifteen distinct vocalizations. One call to be deeply studies is the pant-hoot.

Pant-hoot is a call that can be used for long-distance communication across the territory, and can be used more flexibly in other social contexts. These calls can begin with low-pitched, breathy hoots in a sequence that quickly builds with the addition of hoots of the increasingly higher frequency pants produced by emitting breaths and the help of the chimpanzee's diaphragm, vocal cords, and a large throat sac that contributes to the amplitude of the call. Finally, the pant-hoot reaches a tremendous acme with a loud, high-pitched scream that is often associated with drumming or slapping against trees other hard surfaces in the nearby area. Pant-hoots are the only chimpanzee vocalization yet with this distinctive pattern, including acoustic features that allow other chimpanzees to recognize the individual caller among other things. Both males and females use pant-hoots as an expression of general arousal and excitement during interactions within the groups, but pant-hoots can also provide significant information to distant receivers as well. Pant-hoots are also used to stay in contact with other specific chimpanzees, indicating the location of allies, family members, or nearby strangers because they can carry through the forest canopy.

In addition to providing individual identities, the vocalizations appear to contain both genetic and social learning components. This was revealed when researchers studied the vocalizations of large populations of males from distant chimpanzee communities. Only male calls were recorded because vocalizations differ based on age and gender. Because of the long distance between the two groups' habitats, each was likely isolated from another. Therefore, something else must have been contributing to the different dialects and overlapping components. Several possible conclusions could hence be drawn. Two of the possible conclusions drawn are the possibility of genetic differences between the two populations that are affecting the calls or that the differences in their habitats were impacting the transmission of pant-hoots. A forest environment like Kibale can cause the sound to scatter, but calls are carried better if they are made at lower frequencies and rates. If chimpanzees live in more open areas such as the Mahale community that was studied, calls are not capable of the same level of interference, and calls with a higher rate and frequency would be necessary to ensure that the pant-hoot was heard at a long distance.

The study also revealed that the Uganda site at Kibale had a much higher density of other primates within the same habitat, providing another factor that can contribute to degrading a long-distance call. Even differences in body size between the to groups may add to subtle differences in sound production, since the Mahale chimps are smaller than the chimpanzees living in Kibale. It seems that all these factors might impact the pant-hoots produced by the different communities including the habitat acoustics, body size, and the overall sound environment, respecting the population density of all other species living within the area. The communication of chimpanzees is indeed complex and isn't that far from Bigfoot when it comes to intelligence. There is a single major difference between the two species however: social interactions. Chimpanzees were capable of evolving such complex communication skills because they are generally very social animals. On the other hand, Bigfoot is rather solitary and is rarely ever seen with other individuals. The solitary life of Bigfoot means that it does not require such a complex diversity of vocalizations as seen in chimpanzees. Hence, it is odd if Bigfoot would evolve such an unnecessary feature for a solitary species as itself.

Math also supports the fact that Bigfoot, as a giant ape, is unlikely to occur. Statistics have shown that there were 3,313 reports of Bigfoot in 92 years. According to several sources, there is an estimated trend of Bigfoot reports throughout the years 1921 and 2012 that can be calculated. The trend begins to increase in the year 1950, increasing dramatically in the years 1966 and 1991. Using the linear equation on the amount of reports, then the estimated number of reports by the year 2020 is roughly 4,300. According to researchers, 60% of people would more likely to lie than tell the truth. Giving each person from the large number of people a chance of 40% on telling the truth and using the math of probability would give the percentage of all witnesses telling the truth below 0.02%. By 2020 where the chance of all Bigfoot is the lowest, the chance of Bigfoot actually existing according to witness records is 0.0093%, or a 1 per 10,753 chance of occurring. At this rate, a person is over a hundred times more likely to die from a drug or medical overdose and 16.7 more likely to die from a car accident than the chance of Bigfoot's existence being proven by witness accounts alone.

Not only does the math of probability support the unlikelihood of Bigfoot's existence, but so does population density. The population of Bigfoot in North America is estimated to be 100,000 and as previously mentioned in the start of the chapter, the area of North America is roughly 24.71 million square kilometers. Analysis has shown that around 47% of the United States is uninhabited by humans. Assuming that this applies to the rest of America, then all 100,000 of the Bigfoot population would live in an area of 11,613,700 square kilometers. This would place the population of Bigfoot at a population density of 0.086 per square kilometer at the maximum. Taken the entire area of North America as Bigfoot is found in most, if not all, states of America, then population density is less than half at 0.04 per square kilometer. Taken these two numbers, each Bigfoot should be found in every 11.6 to 24.7 square kilometers, or a length of 3.4 to 5 kilometers. This means that there are either way too little reports or the entire behavior of the cryptid is false. As both of these act like two of the strongest pillars in the evidence used to support those that believe in Bigfoot's existence, then Bigfoot has clearly been debunked by both science and math. While these might not be enough to persuade believers that Bigfoot is indeed the grolar bear, there are DNA tests and carcasses to put an end to the legends of the cryptid.

Bigfootologists have supposedly collected hair samples of Bigfoot from throughout North America, but none proved to contain the DNA of a species unknown to science, let alone a giant ape. Grolar bears on the other hand, have been proven to be real by DNA testing; being estimated to evolve in the Pleistocene epoch. Moreover, there has been no genuine Bigfoot carcass yet to be discovered. This raises the question of where the dead bodies could disappear into, which is nearly impossible to hide both in the form of fossils and decomposing remains. Even the footprints of Bigfoot have been proven to be those of bears.

With all the facts laid out however, the question of the reliability of the man's claims in the YouTube video that shows the head of a Bigfoot that was shot by his father in 1953 is still left unanswered. The Bigfoot's head was estimated to be roughly fifty kilograms. If so, then the supposed Bigfoot specimen would weigh in a colossal 625 kilograms, which is equal to roughly 140% of the upper estimated weight of Bigfoot if given the right proportions. Assuming that the Bigfoot specimen weighs as much as the upper estimated weight, then the head would be around 11% of the Bigfoot's mass, or one ninth of the total body mass. This means that the only remaining possibility for the supposed Bigfoot to be genuine is that the head of the Bigfoot is relatively larger than every single other Bigfoot when comparing it to the rest of the body.

This would therefore cause the captured Bigfoot either a completely different, yet undecipherable, species than the cryptid, or a hoax. The reason behind this can be used when comparing the two most popular species of hominids: modern humans and Neanderthals. Modern humans are more gracefully built than the Neanderthals, which are built to be short and stocky in order to adapt to the colder climates. After close examination on the build of the torso of both modern humans and Neanderthals, it has been calculated that Neanderthals had a relatively larger torso that is roughly 25% larger than modern humans. Although this is drastically larger than the 2% of the face's increased size, it still shows that even a low percentage of differences in the proportions of the body would cause drastic changes in the body's function, in an effect known as the '*ripple effect*'.

Nevertheless, with the evidence clearly not adding up to the specimen being a genuine Bigfoot, it is safe to state that Bigfoot is a hoax. With all types of scientific and mathematical evidence supporting the fact that the Bigfoot is actually a misinterpreted grolar bear, and that video evidence of the supposed cryptid is either a hoax or simply too vague to identify the creature caught on camera, hence believing the existence of Bigfoot.

CHAPTER II: THE HIMALAYAN YETI

The Yeti, also known as the Abominable Snowman, is the giant ape of the Himalayas. This cryptid is treated quite differently than its North American counterpart, with the Himalayan locals treating the Yeti as a spiritual animal. Not only that, but anthropologists have recovered and identified the cranial remains of Gigantopithecus Blacki. Further analysis on the remains shows that the giant ape supposedly went extinct a few thousand years ago near the Himalayan Mountains, being a near exact match to the Yeti's descriptions.

Because of the remote location of the Yeti's habitat, many people would be led to believe that Gigantopithecus would've never went extinct and now exists in the Himalayan mountains as the Yeti. To understand the plausibility of this theory, the ecology, meteorology, and geography of the Himalayas must be studied.

The Himalayas form a barrier between the Plateau of Tibet to the north and the alluvial plains of the Indian subcontinent towards the south. Included in the Himalayas are the highest mountains in the world, with more than 110 peaks rising to elevations of at least 7,300 meters above sea level. The highest mountain is Mount Everest, with an elevation of 8,850 meters. Because of the high peaks of the mountains, the peaks rise into the zone of perpetual snow. For thousands of years the Himalayas have held a profound importance for the people of South Asia as reflected by their literature, mythologies, and religions. Since ancient times, the vast, glaciated heights of the Himalayan mountains attracted the attention of the pilgrim mountaineers of India, coining the Sanskrit name "Himalaya" from "hima" (snow) and "alaya" (abode). In modern times, the Himalayas offer the greatest attraction and greatest challenge to mountaineers throughout the world.

Forming a nearly impassable barrier between the northern border of the Indian subcontinent and the lands towards the north, the ranges are part of a vast mountain belt that stretches halfway through the world, beginning from North Africa to the Pacific Ocean coast of Southeastern Asia. The Himalayas themselves stretch for about 2,500 kilometers (1,550 miles) from west to east between the Pakistani-administrated portion of the Kashmir region and the Tibet Autonomous region of China.

The Himalayas are bordered to the northwest by the mountain ranges of the Hindu Kush and Karakoram. In the north, the Plateau of Tibet borders the Himalayan mountain ranges. From south to north, the width of the Himalayas varies between 200 and 400 kilometers (125 and 250 miles). Their total area accounts to about 595,000 square kilometers, or 230,000 square miles. Although India, Nepal, and Bhutan have the supreme authority over most of the Himalayas, it is also partially occupied by Pakistan and China. In the disputed Kashmir region, Pakistan has authority over roughly 83,900 square kilometers, or 32,400 square miles, of the range lying in the north and west of the "line of control" established between India and Pakistan in 1972. China administers around 36,000 square kilometers, 14,000 square miles, of the Himalayas in the Ladakh district of Kashmir and has claimed its territory at the eastern end of the Himalayas within the Indian state of Arunachal Pradesh.

The Himalayas possess several distinct features, the most characteristic features of which being their soaring heights, steep, jagged peaks, valley, and alpine glaciers that are often massive in size. The topography is deeply cut by erosion, seemingly inscrutable river gorges, complex geologic structure, and the series of elevation zones displaying different ecological associations of flora, fauna, and climate are also some of the most distinct features of the Himalayas. When viewed from the south, the Himalayas would appear as a colossal crescent with the main axis rising above the snow line where snowfields, alpine glaciers, and avalanches feed glaciers in lower valleys that in turn constitute the sources of most of the Himalayan rivers. However, the majority of the Himalayas lies below the snow line. The process that created the Himalayan rages is still active. As the bedrock is lifted, a considerable amount of steam erosion and colossal landslides occur.

The Himalayan ranges can be grouped into four parallel longitudinal mountain belts of varying width, each having distinct physiographic features and geologic history. They are designated from south to north as the Outer Himalayas, the Lesser Himalayas, the Great Himalaya Range, and the Tethys Himalayas. Farther north lie the Trans-Himalayas in Tibet proper. From west to east the Himalayas are divided broadly into three mountainous regions: the western, central, and eastern regions. Over the span of 65 million years, powerful global plate-tectonic forces moved the Earth's crust to form the band of Eurasian mountain ranges that stretch from the Alps to the mountains of southeastern Asia including the Himalayas. Fifty million years ago, the Indian and Eurasian tectonic plates collided, resulting in the formation of the Himalayas and the Tibetan Plateau. The Himalayas are still increasing in size by one centimeter a year as the Indian tectonic plate continues to move northwards towards Asia.

In modern times, the Himalayan population can be classified into three ethnic types: Aryans, Mongoloids, and Negroids. However, the truth about the indigenous inhabitants of the Himalayas is still debatable. One theory states that the first settlement in the Himalayas began around 1500 BC when a warrior tribe by the name of Khasa migrated to its western range. Following Khasa's migration, the Tibeto-Burman people from southeastern Asia to the eastern and central Himalayas in the early part of the millennia. The Tibeto-Burman people in the Himalayas were called the Kiratas. However, if a person were to take into account the Hindu epics and Puranas, then they would deduce that the native inhabitants of the Himalayas were Kinnars, Kulinds, and Kilinds. Darads and Khasas would only migrate later than the native inhabitants of the region. While the exact origins are unknown, the migration within the Himalayas and into it has been proven to happen since the earliest of times. There are plenty of reasons for the migrations in the Himalayas. Some people migrated to the Himalayan mountains in the quest of spirituality while some migrated to test their willpower and endurance. For some other people, the migration was their pursuit of profits and to others; it was the political pressure from their states and countries. Over a period of time, these and other factors combined to give the Himalayas an ethnographically complex population.

If the Himalayan natives could be segregated in terms of their ethnicity in the Himalayas, it would be seen that those living in the higher altitudes, southern side, and those in the northern slopes would belong to the Mongloid ethnicity, which has remained pure because they have a fairly low contact with outsiders. On the other hand, the middle and lower ranges of the southern slopes are home to mixed and diverse ethnic groups with Aryan, Negroid, and Mongoloid strains. Therefore, this can be attributed to regular migrations, invasions, and conquests in these regions. Segregating the natives based on religion, it can be concluded that Hindus predominantly inhabit the Middle Himalayan and sub-Himalayan valleys. The same applies for the area from eastern Kashmir to Nepal. Muslims are mostly found in the western part of Kashmir, with their culture being similar to that of Afghanis and Iranians. The Greater Himalayan region in the north is mainly dominated by Tibetan Buddhists who are found from Ladakh to northeastern India. In the eastern Himalayan region of India and nearby areas of eastern Bhutan, the culture and faith practiced is similar to those followed by the Yunnan Province of China and Northern Myanmar. In Nepal, both Tibetan and Hindu cultures flourish. As a result, this Himalayan nation possesses a mixed cultural identity.

In the Himalayas, each distinct community and valley has its own sociocultural methods the various challenges of life. This is when the communities of the Himalayas cut off from the rest of the global population. However, the slightly common geographical factors have meant that the distinct communities are similar to one another. Although each person has a different opinion, it is indeed a fact that the physical isolation of the Himalayan people has had the positive result of the preservation of centuries-old knowledge. Another factual aspect of all Himalayan natives and its foothills is the fact that they worship the mountains as their life-giver, preserver, and protector. All the communities living in the Himalayan region are dependent on nature, and are strongly ethnic and religious. However, these aspects might not hold true for those residing in the arid wilderness of the dense forests of the eastern slopes and northern flanks where people are fierce and warrior-like. Generally however, the people living in the Himalayas are peace loving. This can be observed and experienced when exploring the various Himalayan destinations. The warm hospitality of the citizens would show their full respect towards the environment they live in, and that their harsh living conditions do not hinder their spirits or way of enjoying life.

The Himalayan natives are also known for their rich tapestry of traditional knowledge that spreads across medicine, architecture, and agro-forestry. In terms of occupation, Himalayan natives can be categorized as nomadic pastoralists and subsistence farmers depending on their specific location and agro-climatic conditions. The majority of the Himalayan population is able to sustain itself through agriculture and animal husbandry. In the higher regions of the Himalayas, society is more liberal than the more conservative lower ranges. Women in the Himalayas are responsible for all sedimentary activities such as gathering fodder, fuel wood, farming, and cooking. Men on the other hand, are responsible for trade activities or managing the animal herds. There have been some changes on the lifestyle of the Himalayan natives as the transportation system and communication improved in the recent years. Modernization has definitely affected their social and cultural systems in several ways. This would hold especially true for the frontier villages of Garhwali, Himachal, Ladakh, and Kumaon where there has been a sudden increase in trade and tourism.

Tourism might have been a momentum for the development of the Himalayan regions, but it has also caused an imbalance. Consequently, the sanctity of the Himalayas is taking a toll. One of the biggest reasons for this is the air pollution caused by vehicles, drastically increasing in the peak tourist season. The increase in tourism has resulted in the overuse of water resources that are in scarcity in some high-altitude regions of the Himalayas. Another important factor causing an imbalance in the region is noise pollution as tourism in the Himalayas increases annually.

The effects of over tourism are certainly hindering the Himalayan nature as well as the natives. Although the situation is currently still under control, the control might loosen over time. The concerned authorities might need to take drastic steps to stop the environmental damage that the Himalayas are facing.

GIANT APES: THE CULTURALLY DIVERSE CRYPTIDS
SECTION 1: TAXONOMY

The evolution of primates has been discussed in the first chapter only for the evolution to not satisfy the amount of evidence required to prove the existence of the North American Bigfoot as a giant ape. In the Himalayan Yeti however, it is more possible for this cryptid to be an ape due to its remote habitat and existence near the habitat of the renowned Gigantopithecus Blacki. It is possible to go into greater depths into the biology of the Yeti in order to taxonomically place the cryptid on the primate family tree.

The Yeti is described to weigh anywhere between approximately 91 to 181 kilograms, or 200 to 400 pounds. It is also described to be muscular with dark grayish or reddish-brown hair, as well as a relatively short height compared to the North American Bigfoot at a height of 1.83 meters, or 6 feet. This description is the most common form out of the various shapes given to the Yeti, and hence the most reliable. In 2010, hunters located in China caught a strange animal that they referred to as a Yeti. The hairless, quadruped animal was initially described as having features similar to those of bears, only to be correctly identified as a civet that has lost its hair from a disease. This can infer that genetics are a main method in the identification of the Himalayan Yeti.

Just like its North American counterpart, there is a lack of solid evidence supporting the Yeti's existence as most of the evidence supporting it comes from sightings and reports. Over the years however, there are a few pieces of evidence that would help explain the general shape of the Yeti. One of these pieces of evidence comes by Sir Edmund Hillary in 1960. In that year, Sir Edmund Hillary who is known to be the first man to scale Mount Everest was searching for evidence of the Yeti. He discovered what was claimed to be a scalp from a Yeti. However, scientists later proved that the helmet-shaped hide was made from a serow, a Himalayan animal that is similar to a goat. Another example comes over four decades later. In 2007, American TV show host Josh Gates claimed that he had found three mysterious footprints in the snow near a stream in the Himalayas. Locals were skeptical, which suggest that Gates, only being in the area for roughly a week, simply misinterpreted a bear track. Instead of being put in display in a natural history museum, the track has been put in a small display at Disney World.

More interestingly however, was a discovery made in Nepal. A finger long that was believed to be from a Yeti was found in a monastery in Nepal and was examined by researchers at the Edinburgh Zoo in 2011. The finger caused a major controversy between Bigfoot and Yeti believers for decades until DNA analysis proved that the finger was that of a human, which might've come from a monk's corpse.

Nonetheless, these descriptions show that the Yeti is a primate. This hints at the possibility that the Yeti is closely related to Gigantopithecus Blacki. "Gigantopithecus" is a term meaning Giant Ape as referred to the largest species, G. Blacki, that stands up to three meters tall and weighs up to 540 kilograms. Gigantopithecus species were discovered in China, India, and Vietnam from the Messinian period of the Miocene to the Late Ionian of the Pleistocene, or 9 million to 100 thousand years ago. While it isn't confirmed, Gigantopithecus possibly lived to later times. This species of ape is part of the Ponginae family of Hominidae. As the finger of a Yeti can and did get mistaken for a finger of a human, it is safe to state that the Yeti would be found in the Ponginae family. It is clear out of all the theories surrounding the Yeti; it can be placed in the Gigantopithecus genus, which has been mentioned in the previous chapter. However, this is unlikely as discussed in the next section.

SECTION 2: THEORIES

Compared to its North American counterpart, the Himalayan Yeti doesn't have as much theories surrounding it. While the Yeti theories lack in quantity, they make for it in quality. Many of the Bigfoot theories surround conspiracy theories and paranormal activities, which are mere speculation and have little to no chance of occurring.

The first theory about the Yeti is the Gigantopithecus theory. This theory states that Gigantopithecus never went extinct and now lives in the Himalayas where the Yeti can be found. Nonetheless, the difference between the Yeti and Bigfoot is that while the Yeti stayed in the Asian continent, the Bigfoot would've gone to North America and evolved. However, there is yet any evidence to support such a claim and therefore Bigfoot can easily be debunked. Just as it is easy to debunk Bigfoot, this theory of the Yeti can be debunked. While the evolution theory is indeed scientifically plausible compared to most theories, it is all but hypothetical as there is no solid evidence to support such claims. There had been yet no fossil remains that suggests of a mainly bipedal ape that has evolved from a species that has no fossil evidence to have lived longer than what is concluded by paleoanthropologists. Another reason on why this theory is unlikely is the diet of Gigantopithecus.

Based on the dental formula, Gigantopithecus had a diet consisting of fruits and foliage. As the Yeti is one of the few animals to live above the snow line, the amount of fruits and foliage that Gigantopithecus had relied on for 8.9 million years were relatively scarce for apes in their massive size and population when looking at the amount of primary consumers in the Himalayas. While the evolution of the feeding methods is indeed possible, ecology proves that there is a major problem: The Yeti would be considered as an invasive species as the Gigantopithecus is gone extinct a mere 100,000 years ago compared to hundreds of thousands of years if not millions of years spent for animals such as snow leopards, black bears, and tahr to evolve in such harsh conditions. The first major problem to be discussed about Gigantopithecus is the disruption of the food web. A food web is a model that shows the feeding relationships between the species found in an ecosystem. Food webs are better than the food chain in showing the complex feeding relationships in an ecosystem.

As many species of primary consumers such as the Himalayan Tahr and the Yak are dependent on the producers that Gigantopithecus feed on, the Gigantopithecus would then act as a limiting factor to the populations of the herbivorous animals, hence lowering the carrying capacity. Limiting factors are any factors or conditions that limit the growth of a population in any ecosystem. For example, too much or too little of any biotic or abiotic factor such as food, water, or light makes an ecosystem unstable, therefore causing change in the ecosystem. The carrying capacity is the state in which a population can no longer grow. An ecosystem's carrying capacity is different for each population. For example, a Isle Royale supports more moose than wolves. This is due to the fact that the moose is at a lower level in the food pyramid in than the wolf. According to the level of the moose and the wolf in the food pyramid, the amount of energy available to wolves is less than the amount of energy available to the moose. Biotic factors including interactions between populations such as competition and predation act as limiting factors. Abiotic factors such as temperature and the availability of water are also limiting factors.

The appearance of Gigantopithecus acts as a biotic factor in the form of competition. One of the several methods of interaction, competition is the struggle between individuals or different populations for a limited resource. In an ecosystem, competition may occur within the same species or between members of separate species. An example of competition within the same species is the creosote bushes. Creosote bushes compete for the same water supply by producing toxins from the root that prevents other creosote bushes from growing. An example of competition of members of different species is the competition of strangler figs and trees in the tropical rain forests of Indonesia. Strangler figs are vines that compete with trees for water, light, and nutrients by attaching itself to a host tree. As the vine grows, it surrounds and eventually kills the host tree by blocking out sunlight and using up the available water and nutrients. In the case of the Yeti, it would compete with the other herbivorous animals of the Himalayas over the available food. It should be noted however, that competition does not occur between all populations that share the same resources Many populations can coexist in a habitat such as maple trees, beech trees, and birch trees coexisting in the same habitat with enough water, nutrients, and sunlight to meet their needs.

Not only would Gigantopithecus face ecological problems, but it would also face evolutionary problems. While Gigantopithecus might have evolved to the steep angles of the mountains by its bipedal stance, the physical descriptions of the Yeti themselves suggest at a major problem in the evolution of the Yeti: climate.

The physical descriptions of the Yeti suggest that it has evolved in a similar fashion to the better known Bigfoot of North America. To figure out whether the Yeti could survive in such a cold climate with the same features as the North American Bigfoot, there is a comparison that could be used. Arguably the most famous polar predator is the polar bear, which has several features that help in the survival in the harsh climates. The first feature is the polar bears' iconic white fur, which the Yeti clearly lacks. Secondly, polar bears have thick layers of fat and fur for insulation against the cold. While the Yeti does indeed have seemingly thick layers of fur, the fur of the Yeti is closer to that of the orangutan. The Yeti also lacks the thick layers of fat with descriptions rarely ever stating the Yeti being seemingly obese and is far from the general view on the cryptid. Thirdly, polar bears have greasy coats that shed water after swimming. The Yeti does not require this feature, as its habitat is landlocked, so it is quite reasonable that the Yeti does not share this feature. Fourthly, polar bears have large, furry feet to distribute their weight and get a good grip on the ice. Similarly to the previous attribute, the Yeti does not require the feature hence it not appearing in the cryptid. Finally, polar bears evolved a small surface area to volume ratio. In the Yeti however, this is not the case as there is a drastic difference of 80% between the surface area and its volume.

This would suggest that the Yeti would live in lower elevations where the reports are relatively rare. A primate that lives similarly to the description of the Yeti is the Assamese macaque. These monkeys prefer dense forests between 610 meters and 1.83 kilometers above the sea level or 2,000 to 6,000 feet, which is unlike rhesus monkeys that are frequently found near humans. During the winter, Assamese macaques rarely travel lands higher than 1.22 kilometers in elevation, or 4,000 feet. Assamese macaques are slightly larger than rhesus macaques, which makes them the largest macaque species in India. These primates live throughout the Himalayas and although they are generally less aggressive, Assamese macaques still share many behavioral traits with the rhesus macaques. However, there would be major conflict with the natives of the Himalayas and ecology if that was the case, hence the possibility is as low as it can get. Nonetheless, the theory is one of the most scientifically plausible theories about the Yeti.

Another theory about what the Yeti could be is that the Yeti is actually a Neanderthal. While this theory is more common and scientifically plausible in the case of the Russian Almasty, the theory still is strong in the case of the Himalayan Yeti. Nevertheless, there is evidence supporting the Neanderthal theory.

The evidence supporting the Neanderthal theory is the fact that Neanderthals are extremely intelligent and were driven to remote places by Homo Sapiens, modern humans. Unlike the Gigantopithecus however, Neanderthals we're capable of surviving in cold regions as they lived in the Ice Age. Ice ages are the glacial periods of the Earth's history. The Earth has experienced several ice ages, the most famous of which is the Pleistocene Ice Age. Also known as the Quaternary or Pleistocene glaciation, the ice age started 2.6 million years ago. Ever since, the ice has ebbed and flowed several times, reaching its 18,000 years ago when Arctic ice sheets covered much of North America and northern Eurasia. During this ice age, so much water has been locked up as ice that the global sea levels dropped by 100 meters, or 330 feet. Shallow coastal seas dried out, including the Bering Strait between Siberia and Alaska, allowing animals and people to migrate between Asia and America. By 11,000 years ago, rising temperatures had melted much of the ice, filling up the seas once again and led to the extinction of many species. The most dramatic ice age however, may have occurred 650 million years ago. Clues found in rocks suggest that the entire Earth was frozen with the region at the equator being as cold as Antarctica is today. However, the hypothesis named the "Snowball Earth" hypothesis is disputed by many scientists because of the fact that such an event would have had a catastrophic effect on the evolution of life.

Ice ages are caused by continental drifts, which move land into colder regions of the Earth, and by variations in the Earth's orbit around the Sun. The Earth has three orbital cycles around the Sun. First is a 100,000-year cycle where the Earth's orbit changes from circular to elliptical. Second is a 42,000-year cycle that affects the tilt of the Earth's axis. Third is a 25,800-year cycle where the Earth would wobble on its axis. There are times where the three cycles would combine, chilling the Earth enough to cause ice sheets to start growing. Unlike popular belief, the Earth's distance from the Sun throughout the year changes by a mere 3%, hence the amount of solar energy received by the Earth annually changes by 6%. However, the change in the amount of solar energy received annually does not cause the four seasons. Instead, the seasons are caused by the 23.5-degree tilt of the Earth's rotation axis. All in all, ice ages are estimated to occur every 100,000 years.

Using computer stimulations, a Japanese, Swiss, and American team including an emeritus professor of physical climatology of ETH Zurich named Heinz Blatter, has managed to demonstrate that the interchange between the ice ages and warm periods depends heavily on the alternating influence of the continental ice sheets and climate. Heinz Blatter explains the feedback effect as the following statement,

"If an entire continent is covered in a layer of ice that is 2,000 to 3,000 meters thick, the topography is completely different. This and the different albedo of glacial ice compared to ice-free earth lead to considerable changes in the surface temperature and the air circulation in the atmosphere."

As the scientists from Tokyo University, ETH Zurich, and Columbia University demonstrated in their paper which is published in the journal Nature, these feedback effects between the Earth and the climate occur on top of other known processes. It has long been known that the climate is highly influenced by insolation on the long-term time scales. Because Earth's rotation and its orbit around the sun slightly change periodically, the insolation also varies. If the variation is examined in detail, different overlapping cycles of roughly 20,000, 40,000, and 100,000 years are recognizable. Given the fact that the 100,000-year insolation cycle is relatively, scientists are unable to easily explain the prominent cycle of ice ages with only this information. With the aid of feedback effects however, it has recently been possible.

The researchers obtained their results from a comprehensive computer model where they combined a simulation of an ice-sheet with an existing climate model, enabling them to calculate the glaciation of the northern hemisphere during the last 400,000 years. Not only does the model take the astronomical parameter values, ground topography, and the physical flow properties of glacial ice into account, but so does it take account of the climate and feedback effects. Using the model, the researchers were able to explain the reason behind the gradual beginnings of ice ages and their relatively sudden stop where the ice-age ice masses accumulate over tens of thousands of years and recede within the span of a few thousand years. The reason behind this is the surface temperature, precipitation, and size of the ice sheet that determine whether they will grow and shrink. As Blatter easily explains it, the larger the ice sheet is, the colder the climate would have to be to preserve it.

Author Name: Alwaleed Alghanim

While Neanderthals did live in the Pleistocene glaciation, they never reached to its peak 18,000 years ago as they went extinct 40,000 years ago. As the peak of the Pleistocene glaciation contains some of the richest deposits, with many thousands of specimens being discovered to a single species, the fact that there aren't any fossils of Neanderthals during that time proves that the Neanderthals would've been long gone. Even Heinz Blatter's statement about the effects of ice ages on the climate suggests the high probability that Homo Sapiens isn't the only threat to the Neanderthals that drove them to extinction.

There is also the fact that Neanderthals might be able to grow from 1.6-1.8 meters to 2-3 meters, as their diet would make them. Although these act a small part of the evidence supporting the theory, these two main claims actually gives enough evidence to debunk the theory. Studies have shown that marine animals have grown by 150-fold in a span of 542 million years. Assuming that Neanderthals grew at a similar pace, then ever since Neanderthals went extinct roughly 40,000 years ago, they would have merely grown two centimeters, 15 times less than the growth that would've been necessary when taking the lower estimates of the sizes of both the Yeti and the Neanderthals. Other than size, the hair and its coloration of both the Yeti and the Neanderthals also prove that the theory's "scientific analysis" is neither accurate nor thorough. The Yeti has a lot more hair than the Neanderthals, with the comparison between the two being similar to Bigfoot and Leif Erikson. While both Leif Erikson's men and Neanderthals indeed have more hair than an average person, Bigfoot and the Yeti possess relatively more hair than any human. Not only that, but genetic traits also debunk the Neanderthal theory. Brown and black hair are two of the most common hair colors found in humans and therefore brown and black hair colors must also be common in Neanderthals. The reddish brown hair described in the Yeti is actually auburn, a variety of red hair. Brown and black hair are dominant traits while auburn hair is a recessive trait. This means that it is very unlikely, if possible in the first place, for Neanderthals' hair to turn into that of the Yeti. To put in additional evidence in order to cover any potential holes in the opposition against the Neanderthal, the biochemistry of hair pigmentation found in humans is involved.

In all humans, two types of pigments, or melanin, called eumelanins and pheomelanins, cause the differentiation in the hair color. Eumelanins are the black and dark brown pigments while pheomelanins are the red and blonde pigments. Combinations of the two basic biochemical structures that have been mentioned cause the different colors of the hair in different people. By mixing the two types of pigments together in different concentrations, many different shades of hair color are made.

Eumelanins are very strong and stable proteins that are made from tyrosine. The eumelanin biochemical structure is formed by the amino acid named tyrosine into dopa and dopamine as well as connecting several of the molecules together to form eumelanin. In this process, the key enzyme is tyrosinase. The amount of eumelanin formed increases as the activity of tyrosinase increases. An increase in the activity of tyrosinase results in more pigment production, hence causing a darker hair color. As a person gets older, the activity of tyrosinase increases, being most active in middle age. Afterwards, the tyrosinase activity decreases. Tyrosinase activity is not the only biochemical mechanism by which the shade of hair color is regulated. Several factors interact with tyrosinase to help regulate the production of eumelanin. Additionally, another key limiting factor in hair color is the availability of the raw tyrosine ingredient. A lack of the available of tyrosine means that tyrosinase enzyme would make eumelanin at full capacity. Pheomelanins are made from the same tyrosine that makes eumelanins and the process is the same with tyrosinase playing a major role. Pheomelanins are produced when an intermediate product in the eumelanin production pathway interacts with an amino acid called cysteine. The interaction results in the formation of a pheomelanin molecule, which contains sulfur from the cysteine. Pheomelanin molecules are yellow to orange in color. The more interaction there is between dopaquinone and cysteine, the more yellow and orange pigments are produced.

Therefore, people with darker hair have relatively more eumelanin production. On the other hand, people with natural red hair produce more pheomelanin than eumelanin. Because people with red hair are less able to produce eumelanin pigments, their skin is typically pale and burns easily with sun exposure. A study made by Borges in 2001 analyzed the amount of eumelanin and pheomelanin in human hair has suggested that black hair has approximately 99% eumelanin and 1% pheomelanin, brown and blond hair contain 95% eumelanin and 5% pheomelanin, and red hair contains 67% eumelanin and 33% pheomelanin. Although this proves that people with dark hair may still produce pheomelanin, the eumelanin pigments largely mask it and therefore not much can be seen. However, pheomelanin pigments are believed to cause the warm, golden, or auburn tones found in some types of brown hair. Therefore, the Yeti is actually considered to have brown hair with auburn tones rather than red. This means that the hair is actually still suited to the heat.

The problem with this however, is the fact that the hair color of the Yeti is not common enough to be the typical hair color of the Yeti. Assuming that the Yeti has the same amount of reports as Bigfoot, then out of all 3,400 sightings, 2,550 of those 3,400 sightings would belong to specimens with dark brown or black hair. 816 of the 3,400 sightings would have blond or light brown hair, and a mere 34 of those sightings would describe red hair. According to this calculation, there is a major flaw in the theory. So overall, genetics and ecology prove that the Neanderthal theory is false.

An explanation for the sighting of the Yeti is the migraines caused by barometric pressure. Barometric pressure is the method that scientists use to measure the atmospheric pressure or weight of the air where it presses on the surface of the Earth. The barometric pressure affects the weather by causing changes to the way that air currents move around the Earth. A device called a barometer is used to identify the pressure, which is useful in forecasting incoming weather changes. High barometric pressure is usually linked to clear, sunny weather while low pressures forms the perfect conditions for the development of clouds and moisture.

As to why the weather conditions cause migraines, there are several theories that exist but with no conclusive evidence to provide an accepted answer. When the barometric reading indicates low pressure, then the weight of the air pressing inwards from the atmosphere is lighter than it could be. The sinuses of humans are filled with air, creating an outward pressure that is countered by the atmospheric pressure. If the inward and outward pressures were unevenly matched, the pressure inside the sinuses would cause them to become distended, particularly in patients with congestion or a blocked nose. In some instances, the change in pressure simply happens at the same time as other weather-related triggers, including the changes in the weather, which cause an imbalance in brain chemicals.

Headaches are effects caused by the barometric pressure in both non-migraine sufferers as well as migraine patients. The headaches are generally experienced on both sides of the head simultaneously, the definition of bilateral. However, a barometric pressure migraine is more frequently felt in just one side of the head. A migraine that is triggered by barometric pressure changes usually lasts an average of two dozen hours, and in some instances it can run up to thrice that span of time at six dozen hours. Apart from the debilitating and severe pain, symptoms of a barometric migraine include the following: Firstly, the person would feel nausea and stomach pains, which are sometimes accompanied by vomiting and diarrhea. Secondly, pain around one or both temples are capable to also affect the eyes, ears, forehead, or the back of the head. Thirdly, there would be feelings of depression and changes in perception of things. Fourthly is the increased sensitivity to light or the development of an aura, which may last for several hours. Fifthly, there would be numbness and tingling in the face, head, and neck, which can also spread to the arms and legs. Finally, waves of pain throb in time with the patient's heartbeat.

Studies have shown that approximately 12% of the population suffers from migraine headaches, and weather conditions are one of the top three most common triggers. Patients with barometric sensitivity report that the pain often begins well ahead of the changing weather patterns. This means that the patient could develop a headache when the weather is visually perfect, only to discover the reason of the pain after a day or two when the clouds finally move in. By the end of October 2017, the population is roughly 7.578 billion. This means that 909.4 million people have migraine headaches, which is over 267,470 times the amount of reports. Therefore, it is indeed possible that this might cause hallucinations that may be connected to the sightings of the Yeti. While this is all hypothetical and less likely to scientifically prove what the Yeti could be, psychology also proves that the brain playing tricks on the person's eyes creates the Yeti.

A person receives their information about the external world by the sensory organs, one of which being the eyes. Specialized nerve cells that evolved to be excited by particular external stimuli such as light or sound provide the raw material in the person's sense experience. The stimulated receptor cells which are also called afferent neurons, set up a chain reaction that excites neighboring cells to create a neural pathway along 'connector neurons' to the central nervous system and the brain. Signals from the brain travel on similar pathways to afferent neurons, also called motor or effector neurons, to stimulate muscles and control bodily movement. This process can be further explained in the areas of the brain.

In the brain, the electrochemical 'messages' sent from the sensory organs are experienced as sensation. Each of the senses sends a different kind of signal, and different areas of the brain, which are analogous to the different sensory organs, process the different kinds of signals. For example, information sent by the eyes is received by the primary visual cortex, located at the rear of the brain and is mainly responsible for vision. Afterwards, the information goes to the neighboring visual association cortex where it is analyzed. To relate this with the Yeti, these basics can be linked with cognitive psychology.

Author Name: Alwaleed Alghanim

In recent times, the term 'cognitive psychology' is associated with the approach to psychology that became predominated after World War II, focusing on mental processes rather than behavior. However, psychologists had set out to the study of how the brain works. Although behaviorism, which dismissed cognitive processes as unobservable and irrelevant, dominated the field of psychology in the USA, German psychologists continued to explore methods to examine the cognitive activities. Hermann Ebbinghaus and Wilhelm Wundt laid the foundations for a scientific study of memory and perception, and they later on laid the foundation of the Gestalt psychology, providing a comprehensive explanation of mental processes that countered behaviorism's emphasis on conditioning. In the 1950s, the cognitive revolution emerged, a movement largely influenced by advances in information and computer science, hence formally starting the study of cognitive psychology.

In cognitive psychology, there are two types of memory storage: short and long-term memory. Short-term memory, or STM, holds information for only a matter of seconds and has a limited capacity. On the other hand, long-term memory, or LTM, can store unlimited amounts of information indefinitely. Short-term memory deals with information that is needed to use immediately, while long-term memory stores anything needed for future use. Most psychologists recognize the dual-store mode of memory, but there is a disagreement as to the exact roles of short and long-term memories, their connection, and whether the two types of memory storage are in fact separate systems. Nonetheless, recalling an incident with the Yeti is considered a long-term memory.

When it comes to long-term memory, the human brain is not entirely accurate. Cognitive psychology's approach to the study of memory was largely based on the analogy of information storage and retrieval. Cognitive psychology's approach was also based on the connection between the two processes. Endel Tulving explained memories as being organized by the human mind into categories, which are put into different stores, showing that these could be recalled by 'jogging the memory' with a cue. Tulving also described remembering as a form of 'mental time travel' taking the person back to the time and circumstances when the memory was stored. The British psychologist Alan Baddeley, who showed that divers would recall memories better in the circumstances in which the divers have learned, later took up this idea. Even later, Gordon H. Bower discovered that a person's memories are cue-dependent as well as mood-dependent where the person's emotions when storing and retrieving memories affect their recall.

For instance, a person in a bad mood would have an easier time recalling bad memories than happy memories.

There are several methods in which memory can be inaccurate. Daniel Schacter has described the seven major methods in which recalling memory can go wrong: transience, absentmindedness, blocking, misattribution, suggestibility, bias, and persistence.

The first method is transience. Transience is the fading of memories with time, especially if the memories are not regularly accessed. Second is absentmindedness. Absentmindedness is caused by flawed storage, which is probably due to the lack of concentration at the time of storing the memory, allowing the mind to categorize the information received as unimportant. Thirdly, the sin of blocking is the occurrence in which a person cannot retrieve a memory because another memory is getting in its way, which leads to what is described as the 'tip of the tongue' syndrome. Fourthly, misattribution is the retrieval process that is flawed, therefore recalling the information would be correct, but the source can be misattributed. Fifthly, suggestibility is the distortion by the cues that trigger recollection. Sixthly, bias is the distortion by the cues that trigger the thoughts and feeling the person had during the memory. Lastly, the sin of persistence is the inability to get rid of a memory.

With research showing that around 67% of information is lost in a span of two dozen hours, a person would have faced any of these 'seven sins' during the encounters if not have simply been a hoax or misidentification. All in all, the scientific explanations of all three theories suggest that the Yeti is most likely a hoax.

CHAPTER III: THE SOUTH AMERICAN MAPINGUARI

Mapinguari lives in the dense jungles and forests of South America. It can be hence concluded that the Mapinguari lives in Chile, Brazil, parts of Paraguay, northern Argentina, Peru, Surinam, and parts of Guyana. As the area takes up much of South America, the geological history of the continent might shed some light to what the creature could possibly be.

South America is the fourth largest continent with an area of around 17.8 million square kilometers, or 6.9 million square miles. Covering most of the western length of the continent is the Andes mountain range, which includes the highest peak in the Americas, Mount Aconcagua. Mount Aconcagua's peak reaches a height of 6,962 meters. East of the Andes is the colossal Amazon river basin with an area of 7 million square kilometers, 2.7 million square miles, mostly covered in tropical rainforest. Also located in South America is the tallest waterfall in the world called Angel Falls. Found on the Churún River in the Guiana Highlands of Venezuela, Angel Falls' highest point reaches an astonishing height of 979 meters, or 3,212 feet. This waterfall was unknown for the most part outside of the surrounding region until the year 1933, when US aviator Jimmie Angel noticed the waterfall as he flew past, hence the origins of the name Angel Falls. Lastly, the Amazon River is the second longest river in the world by length. The Amazon River is 6,400 kilometers long and has the largest water-flow out of any river with an average water discharge greater than that of the Yangtze, the Mississippi, the Yenisei, the Yellow, the Ob, the Parana, and the Congo Rivers combined. Based on the description of the Mapinguari's habitats and the rich ecosystems of South America, it can be concluded that the ideal habitat for Mapinguari would be the Amazon rainforest.

The Amazon spans across eight rapidly countries which are Brazil, Bolivia, Peru, Ecuador, Colombia, Venezuela, Guyana, Suriname, and French Guiana. The Amazon contains 10% of all known species in an area of 5,665,600 square kilometers, or 1.4 billion acres, of dense forests. There is roughly 6,600 kilometers, or 4,100 miles, of winding rivers located in the Amazon. Nevertheless, the colossal area of the Amazon causes many theories about the Mapinguari as the area and biodiversity helps in increasing the amount of scientific evidence.

SECTION 1: BIODIVERSITY

By definition, biodiversity is the variety of plants and animal whether in the world or in a particular habitat. In the Amazon, the biodiversity is so rich that a single bush might have more species of ants than the British Isles. In every 0.01 sq. kilometers, or a single hectare, of the Amazon could hold over half a thousand species of trees and a single park can have more than 1,400 butterfly species.

In the Amazon, there are over 40,000 plant species, 16,000 tree species, 5,600 fish species, 1,300 bird species, 1,000 amphibian species, 430 mammal species, and over 400 species of reptiles. In total, there are over 64,730 species in the Amazon. While there are indeed many species of animals in the Amazon, all species of animals are heavily by the amount of insect species. In fact, insects form over 90% of all animal species found in the Amazon. The Amazon rainforest is estimated to have over 2.5 million species of insects, only a fraction of which has been discovered. Some scientists have estimated that 30% of the animal biomass of the Amazon Basin comes from ants. As a matter of fact, a single sq. mile, or 2.6 sq. kilometers, the forest often accommodates more than 50,000 species of insects.

The Amazon has around 1,100 tributaries with 17 of which being over 1,600 kilometers, or 1,000 miles. In the Amazon, many fishes are important dispersers of tree seeds such as the species Tambaqui. Not only that, but also several traditionally saltwater organisms have adapted to the freshwater conditions found in the Amazon such as stingrays, dolphins, and sponges.

Many birds found in the Amazon are migrants from the north or the south, whether to stay in for the winter or passing through the rainforest at certain times of the year. The world's rarest bird is the Spix's macaw. It has always been rare as it is limited to palm groves and river edges in a small area near the center of Brazil. However, deforestation, importation of Africanized bees - which took the macaws' tree hollows-, and the over-collection of Spix's macaw for the hobbyists caused the macaw's demise. Fortunately however, conservationists are working on restoring a wild population using captive animals.

In the Amazon rainforest, frogs are overwhelmingly the most abundant amphibians with more than a thousand species of frogs located in the Amazon Basin. Unlike temperate frogs that are mostly limited to habitats near a body of water, tropical frogs are most abundant in the trees with a relatively few found near bodies of water on the forest floor. The reason for the occurrence is that frogs must always keep their skin moist since half of their respiration is carried out through their skin. The rainforest's high humidity and frequent rainstorms gives tropical frogs much more freedom to move into the trees and escape the numerous predators in the rainforest waters. Another difference between tropical and temperate frogs is that while the majority of temperate frogs lay their eggs in water, the majority of rainforest species place eggs in vegetation or lay them in the ground. By leaving the water, frogs avoid egg-predators such as fish, shrimp, aquatic insects, and insect larvae.

Among the best known of the Amazon amphibians are the poison dart frogs, which are small frogs with brilliant colors. The slow-moving frogs produce powerful toxins from glands located on their backs and use their color to proclaim their toxic composition to potential predators. Poison dart frogs derive their toxins from the invertebrates that they eat. Another unique amphibian found in the Amazon is Atretochoana eiselti, also known by its nickname, "the penis snake", which was nicknamed for its resemblance to the male penis. This species is actually the world's largest caecilian, a limbless amphibian that resembles an earthworm. Outstandingly, this species can reach up to lengths of 81 centimeters, or 32 inches.

Just as impressive as the amphibians, yet way less revolting than the caecilians, are the mammals. The majority of the mammal species found in the Amazon are bats and rodents. Unsurprisingly, the largest rodent in the world is located in the Amazon. The rodent is called the capybara, which can weigh up to 91 kilograms, or 200 pounds. A group with slightly less common inhabitants in the Amazon is the Edentate group. The Edentate group is a group that includes sloths, anteaters, and armadillos and only exists in the New World. Much rarer than all of the previously mentioned mammals are the freshwater dolphins. In all of the Amazon River, only two species of freshwater dolphins have been discovered.

Finally, there are reptiles. There are more than 450 species of reptiles found in the Amazon Basin, including lizards, snakes, turtles, tortoises, and caiman. Reptiles are actually an important food source for the people in the Amazon, and many reptile species are illegally collected and exported for the international pet trade. After drugs, diamonds, and weapons, live animals are the fourth largest commodities in the smuggling industry.

Because of its rich biodiversity, the Amazon rainforest has many famous animals. Three of the most famous fish are the piranha, electric eel, and the giant catfish. There are multiple famous reptiles and amphibians, some of which are anacondas and boa constrictors, venomous snakes, caiman, frogs, lizards, and freshwater turtles. Many famous birds are located in the Amazon such as macaws and other parrots, toucans, parakeets, hummingbirds, and woodpeckers. Finally, some of the most famous mammals are the big cats, the capybara, and the multiple species of sloths, several species of monkeys, and the Amazon River pink dolphin.

The richness of the Amazon's biodiversity may be a huge factor in the scientific explanation of the Mapinguari, but the biodiversity is hugely connected to the ecology of the Amazon. There are several aspects of ecology that help in explaining the biodiversity of the Amazon as well as hints as to what the Mapinguari could be. Therefore, ecology plays a more major role than the biodiversity, as it will be discussed in the next section.

SECTION 2: ECOLOGY

The Amazon is a tropical rain forest, which is a type of biome. As previously stated, biomes are large geographic areas that are similar in climate with similar types of plants and animals. Tropical rain forests are located near the equator, where the weather is warm all year at around 25 degrees Celsius. Out of all the land biomes, tropical rain forests are the wettest, receiving an annual rainfall of 250 to 400 centimeters, or around 8 to 13 feet. The trees in the tropical rain forest tend to have their leaves throughout the year, providing an advantage because the soil is poor in nutrients. High temperatures cause materials to break down, but there are so many plants that the nutrients get used up just as quickly.

In all of the ecosystems, there are carrying capacities, the maximum number of individuals than an ecosystem can support. There are multiple factors that limit the growth of a population in an ecosystem called limiting factors. Limiting factors can be separated into biotic and abiotic factors. Biotic factors are the factors that include the interactions between populations such as competition, predation, and parasitism. Abiotic factors on the other hand, are the factors that include temperature, availability of water or minerals, and exposure to the wind.

Living organisms depend upon an ecosystem for food, air, water, and other things required for survival. In turn, living organisms have a large impact on the ecosystem that they live in. Plants affect other biotic and abiotic parts of ecosystem. As plants are an important source of food, the types of plants found in a particular ecosystem will determine the types of animals able to live in the same ecosystem. Plants can also affect the temperature by blocking sunlight. The roots of plants hold the soil in place. Plants even affect the atmosphere as they take in carbon dioxide and release oxygen.

Animals are also biotic factors that are able to affect an ecosystem. Beavers build dams that change the flow of a river, affecting the surrounding landscape. Large herds of cattle are able to overgraze a grassland ecosystem, causing the soil to erode. In an ocean biome, corals form colossal reefs that provide food and shelter for marine organisms.

Abiotic factors can either be physical or chemical. The physical factors are the factors that can be seen or felt such as the temperature or the amount of water or sunlight. On the other hand, important chemical factors include the minerals and compounds found in the soil. As there are both physical and chemical abiotic factors, the combination of different abiotic factors from both types determines the types of living organisms than an ecosystem can support.

An important abiotic factor in any type of ecosystem is temperature. In a land ecosystem, temperature affects the types of plants that thrive there. In turn, the types of plants that are available for food and shelter determine the types of animals that can thrive in the same ecosystem. Temperature not only affects the plants themselves, but it also directly affects animals. Similarly to plants, animals are sensitive to temperature. Musk oxen for example, are able to survive freezing environments with an average temperature of -40 degrees Celsius, or -40 degrees Fahrenheit, because of their thick coat of fur. The water buffalo meanwhile, is better suited to warm temperatures that can reach 48 degrees Celsius, or 118 degrees Fahrenheit, because of its lighter coat of fur.

Another important abiotic factor is light. Related to temperature, sunlight warms the Earth's surface. Energy from sunlight also supports all types of living organisms on Earth, especially plants. This is because the Sun provides the energy that the plants use to produce food in the process of photosynthesis. Due to the Earth's 23.5-degree tilt, various regions of the world receive different quantities of sunlight. This therefore causes different types of plants to adapt to the amount of sunlight received in their specific ecosystem. For example, plants such as cacti that are found in desert biomes are suited for living in regions with high quantities of sunlight. Meanwhile, mosses and ferns are more suitable to grow on the forest floor where much of the sunlight is blocked by the trees above.

The third abiotic factor to be discussed is soil, a mixture of small rock and mineral particles. Organisms within the soil break down the remains of dead plants and animals. This process of decay is important to supply significant raw materials to the living organisms of an ecosystem. There are multiple types of soil depending on the different ecosystems. The characteristics of the soil in a particular ecosystem affect plant growth. Soils with a lot of decaying, or organic, matter can hold water well and allow air to reach to the roots of plants. Sandy soils usually aren't able to hold water well because the water flows through sandy soils easily. Clay soils on the other hand, do not allow water to flow through easily because of their relatively small and tightly packed particles.

Lastly, the amount of water available to support life is an essential abiotic factor, as all living organisms require water to carry out important natural processes. Plants require water as well as sunlight to carry out the process of photosynthesis. Animals require water to digest food and release the energy stored in the food. Ecosystems that have a lot of water are able to support a large quantity of different species of plants, which can then support many different species of animals. Tropical rain forests, which are the wettest land ecosystems, are the most diverse. Desert ecosystems on the other hand, have far fewer species of living organisms than tropical rain forests. The amount of species in a land ecosystem is therefore connected with the amount of fresh water that is available for the living organisms in the particular ecosystem.

The different types of biotic and abiotic factors indicate that rain forests such as the Amazon are the most bio diverse and densely populated land biomes on Earth. As previously stated about the Amazon's biodiversity, all of the facts are supported by the Amazon rain forests and therefore the biotic and abiotic factors also affect the types of animals found in the Amazon rain forest. However, there has been a lot of speculation circulating the existence of the Mapinguari using this piece of information.

SECTION 3: THEORIES

The Mapinguari is described to resemble either an ape or a giant ground sloth with red hair, long arms, powerful claws capable of tearing apart palm trees and rip out the tongues of cattle, a sloping back, backwards feet that supposedly make a bottle-shaped footprint, and stands up to 1.83 meters, or 6 feet, tall when it stood on its hind limbs with a bear-like stance. The Mapinguari gave off a decomposing stench and emitted a frightening shriek. It has also been reported to be able to move slowly and stealthily through the forest and surprise the unsuspecting locals. Although the Mapinguari is believed to be carnivorous, there are no reports of the cryptids hunting and eating humans.

Legends state that arrows and bullets aren't capable of penetrating Mapinguari's alligator-like hide. In the late 19th century, a paleontologist examined samples of reserved ancient ground sloth skin, revealing hard dermal ossicles, which are small pieces of bone found in the skin of dinosaurs and alligators that protected them from predators. It is indeed possible that such skin would have been impervious to arrows and bullets.

There are currently three main theories surrounding what the Mapinguari might be: A giant primate, a giant ground sloth, or perhaps an unusual giant anteater. The study of animal growth and development might shed some light on the possibility of such an animal to exist in all three theories about the Mapinguari's identity.

Because growth and development are continuous and dynamic processes that require integration of several physiological functions, nutrition, efficiency of metabolism and respiration, hormonal regulation, immune responses, and the physiological status of the animal, diseases, and the maintenance of homeostasis influence them.

Animal growth and development are separated into the processes that occur before birth, which are called prenatal processes, and the processes that occur after birth, which are called postnatal processes. Animals originate from a single cell, the ovum or the egg, which is fertilized by the male spermatozoon, or sperm, resulting in a zygote. Afterwards, the zygote develops in an enclosed environment, either in the uterus or an egg, for a certain amount of time known as the gestation or incubation period. After birth, young animals experience a period of rapid growth and development until the young animals reach maturity. At that point, some processes such as bone elongation stop while other processes such muscle deposition slow down. The maximum size an animal can reach is determined by its genetics, but nutrition and disease influence an animal's capability to reach its potential size.

Prenatal growth and development are broken down into two stages: embryogenesis and organogenesis. Embryogenesis extends from the unification of female and male gametes to the emergence of the embryonic axis and the development of the organ system during the neurula stage. During embryogenesis, the zygote develops into the blastula and the blastula becomes the gastrula. The zygote is a single cell that is repeatedly cleaved in order to form a multi-celled ball known as the morula. Cleavage is a process involving the mitotic division of the original cell into two cells, which in turn divides into four cells, and so on. While the number of cells doubles at each stage of cleavage, individual cells do not grow in size. Therefore, the morula is the same size as the original zygote, although it is made up of several cells called blastomeres. Cleavage continues until the cells if the developing embryos are reduced to the size of cells in the adult animal. The cells of the morula are rearranged to form a hollow sphere filled with fluid in a stage where the embryo is referred to a blastula. The fluid-filled region inside the cell is called the blastocoel.

The blastula goes through a process called gastrulation and becomes a gastrula. Gastrulation involves extensive rearrangement of the blastomeres. On one side of the blastula, the cells move inward to form a two-layered embryo. In the two-layered embryo, there is the ectoderm layer, which is the outer layer, and the endoderm, which is the inner layer. In between the two layers, a third layer is formed called the mesoderm. The cavity that forms within the gastrula is known as the primitive gut, which later develops into the digestive system. All of the tissues and organs are formed from one of the three layers of cells in the gastrula. After the germ layers are established, the cells rearrange and develop into tissues and organs. Cells grow and differentiate during this phase, which is known as organogenesis. The process of organogenesis spans from the neurela stage to birth. Differentiation is distinguished from the neurela stage, as it is the stage where unspecialized embryonic cells develop into specialized cells destined to form specific tissues or organs.

Differentiation starts at the upper surface of he gastrula. Ectodermic cells divide and form the neural plate. A pair of neural folds, or raised edges, appears and gradually combines to form the neural tube. A mass of cells called the neural crest is pinched off the top of the neural tube and then moves to other parts of the embryo to give to both neural and other structures. Eventually, the thickening of the frontal part of the neural tube forms the brain. The remainder of the neural tube becomes the spinal cord.

In the first several weeks after conception, cells differentiate into organs and body structures. Thereafter, the embryo is referred to as a fetus, and the body structures continue to grow and develop until birth. In humans, it takes approximately 56 days to develop the fetus.

Body tissues and organs are formed in a specific sequence with the head forming before the tail and the spinal cord forming before other organs. Some highly differentiated cells, such as the brain and nerve cells, cannot be replaced if they are destroyed after the original number is fixed during the fetal stage. Therefore, nerve cells that are seriously damaged thereafter are not replaced and are usually kept as permanent damage. Muscle cell numbers are also fixed, with the only difference throughout the animal's lifetime being the increase in size. Bone is capable to increase in size to a degree by environmental conditions as long as it does not exceed the genetic potential of the animal.

The period of postnatal growth spans throughout the animal's lifetime. There are three major types of tissues that develop with the growth of the animal, which are muscle, bone, and fat. Muscle fibers are formed from multiple cells called myoblasts. While the animal is still in the prenatal stage, myoblasts fuse to form myotubes, which develop into muscle fibers. Consequently, a single muscle fiber contains multiple nuclei. Because fibers are only formed before the birth of the animal, the postnatal growth of muscles is distinguished by the increase in length and diameter. Muscle fibers are predominantly protein; therefore fiber size is determined by the rate of protein synthesis subtracted by the rate of degradation.

Bone tissue grows both before and after birth. A bone grows in length through the ossification or hardening of the cartilage at each end. After the cartilage on the ends of a bone has completely hardened, the growth of the bone stops. However, bones have the capability of increasing in width and self-repair when broken. Although individual bones reach a mature length and stop elongating, bone tissue is constantly being deposited and resorbed.

Fat tissue is comprised of fat cells and connective tissue with fat cells increasing or decreasing in size depending on the nutritional status of the animal. Two types of fat tissue include white fat, which stores energy, and brown energy, which maintains a constant body temperature. Fat is deposited in four different areas throughout the body of the animal. Fat deposited in the abdominal cavity around the kidneys and pelvic area is called the intra-abdominal fat, which is usually the first fat to be deposited. The largest amount of fat deposited is just under the skin, and is referred to as the subcutaneous fat. Fat deposited between the muscles of animals is called intramuscular fat while the fat deposited within the muscle is called the intramuscular fat. The level of intramuscular fat deposited is referred to as marbling. Intramuscular fat is the last type of fat to be deposited, so animals with high degrees of marbling also have large amounts of fat deposited in other areas in the body.

Deposition of various tissues and the partition of energy for different processes involved in growth and development are regulated by hormones. Some of the important hormones that are involved in growth and development are insulin, growth hormone, Insulin-like Growth Factor 1 (IGF-1), thyroid hormones, glucocorticoids, and the sex steroids.

Insulin stimulates the transport of certain amino acids into muscle tissue and is active in reducing the rate of protein degradation. It also plays a major role in regulating food intake, nutrient storage, and nutrient partitioning.

Growth hormones meanwhile, stimulate protein anabolism in numerous tissues. The effect reflects increased amino acid uptake, increased protein synthesis, and the decreased oxidation of proteins. Growth hormones also enhance the utilization of fat by the stimulation of triglyceride breakdown and oxidation in adipocytes. Additionally, the growth hormone appears to have a direct effect on bone growth by stimulating the differentiation of chondrocytes. Growth hormone plays a role in maintaining blood glucose within a normal range. Growth hormones' major role in the stimulation of body growth is to stimulate the liver and other tissues to secret IGF-1.

IGF-1 stimulates the proliferation of chondrocytes, or cartilage cells, which results in bone growth. It also plays an important role in the metabolism of protein, fat, and carbohydrate. Furthermore, IGF-1 stimulates the differentiation and proliferation of myoblasts, amino acid uptake, and protein synthesis in muscle and other tissues.

Animals require thyroid hormones for normal growth. Deficiencies of thyroxin and triiodothyronine cause reduced growth as a result of deceased muscle synthesis and increased proteolysis. Additionally, thyroid hormones have an important effect on the prenatal development of muscle.

Glucocorticoids restrict growth and induce muscle wasting with different effects on different types of muscle. It is indicated that glucocorticoids also affect metabolic rate and the balance of energy. Androgens, the male sex hormones, have an obvious effect on muscle development and growth in general because male animals tend to grow faster and develop more muscle than the females of the same species. However, estrogens, the female sex hormones, also have significant roles in maximizing growth and are commonly used in artificial growth promotants for both male and female cattle. It is believed that androgens have a more direct on the secretion of other hormones than estrogen because of the androgen receptors located on muscle cells.

Homeostasis is a concept referring to the maintenance of an internal equilibrium that is closely integrated with the growth and development of an animal. Many processes and functions, both voluntary and involuntary, contribute to the maintenance to the state of internal balance, which is controlled by the nervous system by nervous regulation and the endocrine system by the chemical regulation. Homeostasis is maintained at all levels, from individual cells to the entire animal. Maintaining a state of homeostasis requires a high level of interaction between hormonal and nervous activities.

All types of animals have a set genotype that determines their potential for growth. However, their phenotype is affected by environmental factors such as nutrition, disease, parasites, and injuries. Specific genes code for different traits with some traits are influenced by multiple genes. For example, rate of growth is a trait influenced by many genes controlling appetite, tissue deposition, skeletal development, energy expenditure, body composition, and many more traits. The genes for all of the traits add together to produce the growth rate that can be measured.

Based on Kleiber's law, which is the observation that the metabolic rate of the majority of animals is equal to the 0.75th power of the animal's mass, the growth rate necessary for the Mapinguari can be determined. First, the Mapinguari has an upper metabolic rate of 91.1 based on the upper estimates of its mass, which is equal to roughly 410 kilograms. Next, the metabolic rate of the giant anteater, orangutan, and Megatherium is calculated to compare with the Mapinguari. Based on given information, the giant anteater has a metabolism of 15.9, the orangutan has an upper metabolism of 31.6, and Megatherium has a metabolism of 503. Respectively, the three animals have a metabolism equal to 17.5%, 34.7%, and 552.1% of the metabolism found in the Mapinguari.

Also given by the calculations is that the orangutan is the closest animal to the Mapinguari. That fact an also be supported by the descriptions of the cryptid. The descriptions of Mapinguari vary heavily, but it mainly divides into either an animal similar to the giant ground sloths or as its North American counterpart, a giant primate. As previously stated, the giant primates are most likely relatives of the orangutan, so the theory is plausible up to this point.

The giant primate theory can be debunked by the timeline of primate evolution and the known discoveries of primates in the Amazon. Orangutans have evolved 14 million years ago in Asia, but it would be impossible for them to migrate to the Amazon since the two continents have separated from over 170 million years ago during the Jurassic period. Therefore, only convergent evolution can conclude the plausibility of the giant primate theory.

Convergent evolution is generally defines as the evolution of similar traits in separate evolutionary lineages inhabiting similar environments. In evolutionary biology, convergence has played a key role in no less than three ways. First, convergent evolution provides natural replicates that can be used to address general questions in the field that surpasses the limitations of studying single evolutionary events. Second, convergent evolution has infamously confounded studies that attempt to estimate the phylogenetic relationships among species by identifying convergent traits that might uncover the shared derived characters that denote historical relationships among species. Finally, convergence helps in the prediction of long-term evolution.

Similarly to the Mapinguari, orangutans live in tropical rainforests, which supports the plausibility of the convergent evolution theory. However, the adaptations and ecological niches of the two creatures suggest that the possibility of convergent evolution seen between the animals is highly unlikely.

Unlike the Mapinguari, orangutans are omnivorous. However, meat is only consumed in rare occasions, with most of the orangutans' diet consisting of fruit and leaves that are gathered from the trees of rain forest. Mapinguari however, are carnivorous, lacking the adaptations of eating foliage as seen in orangutans. Another difference is lifestyle. Orangutans have adapted long arms to suit their lifestyle where they spend around 90% of their time in the trees. On the other hand, the Mapinguari is well adapted to the ground, similarly to gorillas and humans. Because of the two animals' differences in lifestyle and diet, they have different ecological niches.

Ecological niches of species are the roles that the species play in their ecosystem. Orangutans are known as the gardeners of the forest because of their vital role in seed dispersal and the maintenance of the health of the forest, which is beneficial for both humans and nearby animals. While the ecological niche of the Mapinguari is unclear, all what is known is that if it indeed a real species, it would have a very distinct ecological niche than that of the orangutans.

Of what is known from the Amazon species, the primates are divided into monkeys and lesser apes with little to no evidence supporting the existence of great apes. Therefore, the lack of evidence supporting the overall existence of such great ape as the Mapinguari, and major differences from its supposedly closest relatives from their diet and lifestyle to their ecological niches, convergent evolution seems highly unlikely. Therefore, the Mapinguari is not a giant ape.

The giant anteater theory is a fairly easy theory to debunk. Giant anteaters do not satisfy the descriptions that state of a bipedal animal with a short snout and a carnivorous diet that consists of cattle as giant anteaters are quadruped with long snouts and necks, and an insectivorous diet that consists of ants. Not only that, but giant anteaters aren't even large enough since the Mapinguari is nearly ten times heavier than the giant anteaters and five times as tall. With a lot of differentiation between the two animals, Mapinguari is clearly not a giant anteater.

Finally, the giant ground sloth theory is one of the most confusing theories relating to the Mapinguari and perhaps the most widely accepted amongst believers. Immediately, there seems to be a problem since there is more than one giant ground sloth that lived in South America. Four of the main giant ground sloths in South America are Megatherium, Eremotherium, Thalassocnus, and Glossotherium. Glossotherium, Eremotherium and Megatherium are simply too large to become the Mapinguari as they are 2.4, 6.6, and 9.8 times larger than the Mapinguari by mass respectively. In addition, all three giant ground sloths are herbivorous as opposed to the carnivorous diet of the Mapinguari. The most persuasive evidence that shows that Mapinguari is not a giant ground sloth comes in the form of paleontology.

Megatherium was shown to eat dozens of different types of plants by fossilized dung. It typically walked on all four limbs but studies show that it can also rear up on its hind limbs to reach high branches that it pulled with its claws. When upright, Megatherium was almost twice as tall as an elephant, dwarfing the Mapinguari. Megatherium also possessed blunt teeth to mash plants, the exact opposite of Mapinguari. The last evidence of Megatherium was roughly 10,000 years ago, meaning that the animal has gone extinct. Further evidence suggesting that the Megatherium would have no possibility of surviving today is that evidence of the animal disappeared soon after humans first settled in the Americas, which have indeed been found to hunt other animals to extinction.

It has been discovered that Megatherium lived in woodlands, which are different from rainforests such as the Amazon. In rainforests, there is a greater density of trees where the leaves and branches of trees would often meet or interlock. This causes for the areas where sunlight never reaching the ground fairly common. On the other hand, woodlands have many open spaces with a lower density of trees. With large spaces between trees, sunlight can easily penetrate through. In the case of tropical rainforests, the rainforests are hot and humid with a lot of vegetation. Meanwhile, woodlands are cooler with mostly trees and bushes.

With differences in ecosystems, feeding methods, and size, as well as the evidence supporting that it is impossible for the Megatherium to still be alive today, it is illogical to assume that Mapinguari is indeed a Megatherium. As both Eremotherium and Glossotherium show similar traits and ecosystems, it is clear that Mapinguari is not any of these ground sloths. Yet, Thalassocnus is a bit unique between the ground sloths. Even so, it is clear that it is also not the cryptid.

Thalassocnus is a genus of semi-aquatic to fully aquatic marine sloths that lived in Pliocene Peru. Not only did the species go extinct for millions years, but it has a completely different lifestyle than the Mapinguari. While more recent species of the Thalassocnus genus live completely in the water, there are little to no incidents of the Mapinguari ever swimming in the water. Diet is also a problem as Thalassocnus consists of grazers, whether it was in beaches, shallow waters, or in the deep waters. On the other hand, Mapinguari mainly eats animals such as cattle on land. Thalassocnus is also much smaller than the Mapinguari, reaching only 12.2% the mass of the cryptid and 40% the height. With these clear differences, it is impossible for the Mapinguari to be Thalassocnus. That however, leaves more questions about the cryptid than answers.

The Amazon is a really dense area that gives a lot of possibilities to what the Mapinguari can possibly be. Whether the Mapinguari could be an actual animal that can be discovered in the future, a misidentified creature, or simply a hoax is currently not definitive. However, the evidence does lean towards the fact that the animal is simply a hoax.

CHAPTER IV: THE AFRICAN WATERBOBBEJAN

In South Africa, an animal called the Waterbobbejan has been accused of terrorizing people and even killing a few, as well as all types of terror in their livestock. The types of livestock terror include ripping cattle, goats, chicken, and anything else the Waterbobbejan could catch. The name Waterbobbejan translates to "water baboon" and has been described as being pygmy sized to roughly 2.13 meters or 7 feet. Its fur is described as being either red or as black as scorched earth. There are several eyewitness sightings of the creature from the deep woods to the outskirts of African cities.

Jaffe, the world's preeminent expert of the subject, has been interviewed. In the interview, Jaffe has stated that he has examined a genetic sequence taken from what is claimed to be a tuft of the Waterbobbejan's fur. The sequence came out to be similar to human's sequence with an extra chromosome. The fur itself was as thick as a dog's fur. It has also been claimed by Jaffe that the reason behind the lack of the Waterbobbejan in the fossil record is that there are plenty of gaps in the fossil record where the Waterbobbejan fossils could've been found. In the same interview, Jaffe stated that the natives of a remote village near Tanzania recognized a picture of the Waterbobbejan, which is known as the Agogue in the village. The Waterbobbejan was pictured as an oversized human with a pointed head, unusually long arms, and a thick coat of black fur. However, Jaffe himself has stated that this by itself is not convincing, therefore it does not account as a solid fact. There is a lack of historical documents of the South

African Waterbobbejan; therefore the history of South Africa might give clues about the absence of the documents.

South Africa has had a long history, especially with the official system of apartheid that spanned for the majority of the 20th century. With the apartheid system, the government, which was controlled by the minority white population, enforced segregations between the races in their housing, education, and virtually all aspects of life. As an affect, three nations were created. The first is that of the whites, which consists of people primarily of the British and Dutch ancestry, who struggled for generations to gain their political supremacy, reaching its violent apex with the South African War of 1899 to 1902. Second is that of the blacks, which consists of people such as the San hunter-gatherers of the northwestern desert, the Zulu herders of the eastern plateaus, and the Khoekhoe farmers of the southern Cape regions. Finally, the third nation consists of the mixed-race people and ethnic Asians, which consists of Indians, Malays, Filipinos, and Chinese.

In 1960, 69 people died and 180 people were injured when the police turned their guns on a non-violent demonstration organized by the anti-apartheid group, the Pan Africanist Congress (PAC). The massacre that occurred in the township of Sharpeville triggered a shift to more militant tactics among activists. In 1961, Nelson Mandela became the leader of the military wing of the political party in the African National Congress (ANC), and they began a campaign of sabotage that targeted government installations. Mandela and other members of the African National Congress were arrested in 1962 and later were sentenced to life imprisonment. At his trial, Nelson Mandela spoke of freedom, democracy, and equality for all South Africans. Nelson Mandela's long imprisonment became a subject of growing international condemnation.

The "Homeland System", which appeared in the 1960's, aimed to complete the execution of apartheid by the creation of the independent homelands of the blacks. These regions were impoverished rural areas without real capacities to function as the separate states. During that time, 13% of the country was divided into ten homelands with 80% of the population living in the homelands.

The apartheid system was disdained by much of the world community, and South Africa found itself among the world's pariah states, the subject of economic and cultural boycotts that affected almost every aspect of life, by the mid-1980's. During this era, the South African poet Mongane Wally Serote remarked the following:

"*There is an intense need for self-expression among the oppressed in our country. When I say self-expression I don't mean people saying something about themselves. I mean people making history consciously... We neglect the creativity that has made the people able to survive extreme exploitation and oppression. People have survived extreme racism. It means our people have been creative about their lives."*

It took a new administration to start open a way for change. In 1989, president FW De Klerk was elected as president and lifted the bans made on the ANC and the other opposition groups.. Together with Nelson Mandela, De Klerk managed to attain the change of South Africa to the democratic majority rule under difficult circumstances. They faced different views from both tribal and political factions within the black community and opposition to change from some whites. De Klerk and Mandela's solution was to adopt a unified approach and in 1994, Nelson Mandela was elected as the President of South Africa. Hence, Nelson became the head of the Government of National Unity in which the minority such as vice president Klerk's National Party, were represented.

The new government soon recognized that if unity was to be achieved, between the divided communities, it must take action. The Truth and Reconciliation Commission (TRC) appointed the anti-apartheid campaigner Archbishop Desmond Tutu as its head as an historic attempt to label the violence and human rights abuses of the apartheid era. In a span of three years, the commission heard the testimonies of both the victims and the criminals. Although the program had never been intended as an instrument of punishment, the Truth and Reconciliation Commission bore witness to the suffering of victims and could grant the perpetrators amnesty from possible prosecution. In 1988, the commission published its findings. Atrocities committed by all sections of society were condemned, including the actions of the black vigilante group Mandela United Football Club, which was led by Nelson Mandela's former wife, Winnie.

The ambition of the Truth and Reconciliation Commission healed the wounds of a terrible past by acknowledging the truth. South African will to engage in the process won the nation respect throughout the Earth. In his response to the commission's findings, Nelson Mandela called on the people to celebrate and strengthen their actions as a nation. Throughout this long history, there are many incidents in which South Africans would go near the supposed habitats of the Waterbobbejan in the form of mining.

There is evidence showing that small-scale gold mining had been occurring in South Africa's greenstone belt areas sometime before the emergence of modern gold mining industries, but there is little recorded history that exists prior to the 1830's. The more recent history of gold mining in South Africa started with the mining in the greenstone belts in 1836 in northern Kwazulu-Natal and the development of mines in the greenstone belts of Murchison, Giyani, and Pietersberg. Nearly four decades later in 1875, gold was discovered on the farm Kromdraai located just north of the present day Krugersdorp. This led to the first proclamation of gold in the Witwatersrand region. In 1883, the Pioneer Reef was discovered in Barberton, and the very rich Witwatersrand Main Reef was discovered in 1886. Miners from around the world went to South Africa, and many new companies were established as well as the commencement of the development of the country's large-scale gold mining industry. By 2004, three of the size largest international gold mining companies are South African with 342 tons of gold produced in 2004 alone.

There are five large and publicly listed gold mining companies in South Africa, which are: AngloGold Ashanti, Gold Fields, Harmony, DRDGOLD, and Western Areas. Combined, these five companies accounted for 312 tons of fine gold mined in South Africa in 2004 alone, accounting for 91% for all of the gold mined in South Africa.

AngloGold Ashanti currently has seven mines that are operating in two regions of South Africa. First is the Vaal River region, which is near Klerksdorp in the North West Province, comprising of three mines: Great Noligwa, Kopanag, and Tau Lekoa. Second is the West Wits region near Carletonville which comprises three mines: Mponeng, Savuka, and TauTona. Gold Fields is also located in Johannesburg, South Africa with more operations in Ghana and Australia. Although Harmony's corporate is in Johannesburg, Harmony's headquarters are located in Virginia in the Free State. DRDGOLD is also located in Johannesburg, with the company being listed on the JSE Limited, the London, Australian, and Port Moresby stock exchanges as well as Nasdaq in the USA. By the end of June 2005, the total number of people employed at DRDGOLD's South African operations was 6,390. Finally, Western Areas has a stake of 50% in the South Deep mine together with their joint venture partner Placer Dome. In 2004, the South Deep mine produced 13.4 tons of gold, with 6.7 tons of which were attributable to the Western Areas. As of the 31st of December 2004, Western Areas has employed a total of 4,914 people.

With the rich history that should supposedly have more encounters with the Waterbobbejan, from the apartheid to the gold mining history, the answers to many questions about the subject might be found in the cryptid itself.

GIANT APES: THE CULTURALLY DIVERSE CRYPTIDS
SECTION 1: ZOOLOGY

To study the zoology of the Waterbobbejan, there must be a comparison to another animal. The most plausible animal to be related to the Waterbobbejan is a type of ape. However, there are only monkeys found in South Africa with little to no species of ape.

Rumors have existed about the Waterbobbejan since the 1880's, with a significant sighting occurring in 1965. Two boys saw the animal on the Leeufontein farm between Koster and Swartruggens in the North-West Province, South Africa. There are currently two common possible explanations of the identity of the Waterbobbejan. Firstly, the Waterbobbejan could be the Chacma baboon. Chacma baboons are well known in the area but only grow to about 114 cm or 3'9" in length. Secondly, the Waterbobbejan could be the samango monkey. Although samango monkeys are smaller than Chacma baboons, there is at least one incident where a farmer shot and killed a samango monkey and claiming that it was a Waterbobbejan.

Other than the difference in size, chacma baboons and samango monkeys have a different diet than Waterbobbejan. Chacma baboons are omnivorous, preferring fruits while also eating insects, seeds, grass, smaller vertebrate animals, and fungi. Particularly at the Cape of Good Hope, chacma baboons are also known for taking shellfish and other marine invertebrates. Samango monkeys on the other hand, have a diet consisting of fruits, insects, flowers, and leaves. These diets are different than the diet of the Waterbobbejan, as it is mainly carnivorous with a diet partially consisting of livestock such as cattle.

Nonetheless, there aren't any animals in South Africa that are closely related to the Waterbobbejan. Since Waterbobbejan translates to "Water baboon", chacma baboons are theoretically the closest South African animals to the cryptid. Baboons are the largest types of monkeys, and are extremely intelligent animals.

Baboons are able to convey a range of emotional intensity based upon the repetition of some sounds that are associated with other features such as the baboons' facial expressions or postures. This type of communication is considered as the principal form of the baboons' communication.

Some scientists have estimated the number of the baboons' vocal calls at thirty, which includes a range of screams, barks, grunts, and alarm calls. Nonverbal components also form a part of the baboons' communicative repertoire, including their body postures and facial expressions. The communication among baboons does not always require a vocal component. For example, baboons can display aggression efficiently with what is called the "open-mouth threat". This type of threat consists of raising the baboon's eyebrows and revealing the whites of their eyes before the baring of their teeth. At larger levels of hostility, baboons can make their hair stand up, creating the impression of a bigger body size while also including a threatening vocalization, followed by a strong slap to the ground.

A range of studies on several species of baboon show that baboons have flexible and in some cases, specific types of vocalizations, including some species that refer to the different types of predators, identify the caller, or indicate an animal's emotional state. For example, female chacma baboons have a type of call known as "loud barks" that can vary according to each individual caller, the type of predator, and the social context at the time of the call. By studying the acoustic features of the calls in details, field researchers were able to determine that baboons had a range of variability along a continuum. More tonal barks are given when a baboon wished to maintain contact with its group, or as a contact call when an infant becomes separated from its mother. More noisy versions of the call were given as what is known as "alarm barks" when large predators were seen. Within the alarm call category, there were measurable and significant differences between calls that were made in response of seeing alligators and those produced in response to mammalian carnivores. Both types of alarm calls were distinctly different from the types of contact call heard. There were also quantitative, consistent differences in all types of calls for different individuals.

Nonetheless, the Waterbobbejan aren't as social as baboons, so there are differences between the two animals. While chacma baboons show very advanced types of communication, both verbally and nonverbally, the Waterbobbejan would show more primitive features in communication. However, the communication of Waterbobbejan is comparable to that of orangutans.

While the communication of Waterbobbejan is supposedly similar to that of orangutans, its feeding method is not. In fact, the diet of the Waterbobbejan hints overlaps with the diet of several mammalian carnivores in South Africa, including big cats. Because of the overlapping diet and the links between Waterbobbejan and baboons, there are several theories about the animal.

SECTION 2: THEORIES

The first theory about the Waterbobbejan is the chacma baboon theory. Chacma baboons are commonly found in South Africa, and can reach a length of up to 115 centimeters, or 3'9" - excluding their tails -, and could weigh up to 45 kilograms, or 99 pounds.

Evidence supporting the chacma baboon theory includes the translation of the name Waterbobbejan, which as stated many times before hand is "water baboon". Another piece of evidence is that chacma baboons are the longest monkeys with a diet partially consisting of smaller vertebrate animals. The scientific name for chacma baboons is *Papio ursinus*, which is derived from the French words meaning, "Bear like baboons". This definition suits the Waterbobbejan description, and can be even used as evidence supporting the existence of the cryptid. However, there is a lack of further solid evidence supporting that the Waterbobbejan is a chacma baboon.

While chacma baboons are indeed big, they are over twice as small by mass than the supposed cryptid, which would weigh around 122.5 kilograms, or 270 pounds. Chacma baboons also lack the coal black to reddish brown hair that the Waterbobbejan has. Another piece of evidence that shows that Waterbobbejan is not a type of chacma baboon is their diet as mentioned in Section 1. Waterbobbejan have a carnivorous diet while chacma baboons are opportunistic omnivores. Finally, chacma baboons are extremely social animals, living in groups from four to two hundred members. On the other hand, Waterbobbejan are extremely shy animals that are surprisingly solitary despite its name translating to one of the most social groups of primates.

The second theory about the Waterbobbejan is the samango monkey theory. The best piece of evidence supporting this theory is an incident where a farmer shot and killed a samango monkey, claiming that it is a Waterbobbejan. There are differences in sources however, as some other sources claiming that there multiple similar incidents occurring to other farmers. Other than the overall appearance and habitat however, there are many differences between the Waterbobbejan and the samango monkeys. One of the many differences is the fact that the Waterbobbejan can reach masses over fifty five times than the largest male samango monkeys, which can weigh 9 kilograms. Diet is also an important factor because other than insects, samango monkeys lack the carnivorous diet of the Waterbobbejan. The third difference between the two creatures is social activity. Similarly to chacma baboons, samango monkeys are also social animals, which is the opposite of the Waterbobbejan.

The Waterbobbejan's behavior best suits that of leopards. Leopards are a graceful and powerful species of cats, and are the least social species of the cat family. The leopard is so strong and comfortable in the trees that it often hauls its prey into the branches. By dragging the bodies of large animals upwards, leopards hope to keep their prey from scavengers such as hyenas. Leopards are also capable to hunt from trees, where the spotted coats of the leopards allow them to blend with the leaves until they spring with a deadly pounce. Not only that, but leopards also stalk antelope, deer, and pigs by stealthy movements in the tall grass. When there are human settlements are present, leopards often attack dogs and occasionally attack people. Additionally, leopards are strong swimmers and are comfortable in the water as they sometimes eat fish or crabs.

Female leopards can give birth during any time of the year. Females usually have two grayish cubs with barely visible spots. The mother hides her cubs and moves hem from one safe location to another until the cubs are old enough to start playing and learning to hunt. Cubs live with their mothers for about two years, with leopards being solitary animals otherwise.

Similarly to the illustration of the Waterbobbejan in Jaffe's experience with the village he encountered, leopards can have black fur. The black leopards would often appear to have a solid black color because their spots are hard to distinguish from their fur. Leopards with the black fur are often called black panthers.

However, the behavior and color of the leopards' fur do not explain many other attributes of the Waterbobbejan. Taking in the factor of the Waterbobbejan's name and the common plausible explanations for the identity of the cryptid, the Waterbobbejan is much closer to monkeys such as baboons. Leopards can also potentially be longer than the Waterbobbejan when including its tail, reaching a length of 3.28 meters or 10'9". The Waterbobbejan meanwhile, can reach a height of 2.13 meters, or 7 feet.

Lastly, the color of the black panthers' fur does not suit Jaffe's study on the samples of the black fur of the supposed cryptid.

With fur and body of a primate and the behavior of leopards, there is no solid evidence to support the Waterbobbejan's existence in biology. Moreover, paleontology also disproves the existence of Waterbobbejan in the form of leopards and primates.

Modern leopards have evolved 500,000 years ago, and are the most widespread and adaptable of all big cats. An important factor that has played on the successful and a rapid evolution of modern leopards is their diet.

Twenty-nine published and four unpublished studies of leopard diet that contained the relative prey abundance estimates associated with the leopards were analyzed from thirteen countries in forty one separate spatial locations throughout the distribution of the leopards. A Jacobs' index value was calculated for each prey species in each study and the mean of the values was then tested against a mean of 0 using t or sign tests for preference or avoidance. Leopards have been found to prefer prey within a weight range of 10 to 40 kilograms. Regression plots suggest that the most preferred weight of the leopards' prey is 25 kilograms, while the mean body mass of the significantly preferred prey is 23 kilograms. Leopards prefer prey with a weight that falls within this range, which occurs in small herds, in dense habitat, and afford the hunter minimal risk of injury during capture. As a result, impala, bushbuck, and common duiker are significantly preferred, with chital possibly being preferred with a larger sample size from Asian sites. Species that are out of the preferred weight range, the species restricted to open vegetation, and the species that have sufficient anti-predator strategies. The ratio of mean leopard body mass with that of their preferred prey is less than 1 and might be a reflection of the leopards' solitary hunting strategy.

Leopards are catholic in their use of their habitat, ranging from tropical rainforests and arid savanna to alpine mountains and the edges of urban areas. The highest population density of the leopards is in the riparian zones, illustrating that leopards are capable of living in any location with enough cover and adequately sized prey animals.

Leopards vary morphologically drastically, with adults weighing anywhere between 20 and 90 kilograms. They require between 1.6 and 4.9 kilograms of meat daily to maintain their body mass. To achieve this daily food intake, leopards kill around forty prey items yearly in the Londolozi Game Reserve on the border of Kruger National Park, fifty in Kruger, and sixty in the Serengeti. The leopard's body mass largely exceeds the 21.5

kilograms threshold of obligate vertebrate carnivory. However, leopards' varied body mass may enable them to exist for short periods of time on invertebrates or small vertebrates in areas with a lack of large prey. Therefore, it is not surprising that there are reports of leopards preying on small species such as birds, rodents, hares, catfish, and up to the size of giraffe calves and adult male eland. Leopards also have the broadest diet of the large predators with 92 prey species being recorded in the sub-Saharan Africa, although these records are thought to focus on the 20 to 80 kilogram range.

Leopards are almost entirely solitary, with the females' territories being overlapped by the larger territories of the similarly solitary males. In open habitats, leopards hunt alone at night where their camouflage allows them to stalk exceedingly close to their quarry before initiating a short sprint of up to 120 meters. However, leopards' sprints average in 10.3 ± 1.3 meters in Kaudom at up to 60 kilometers per hour. In the contrary, leopards living in rainforests hunt diurnally with crepuscular peaks by ambushing their prey at fruiting trees and along game trails rather than stalking. Attempts only end with kills in 5% of the leopards' hunts in the Serengeti, 16% of hunts in Kruger, and 38% of hunts in Kaudom. Furthermore, between 5 and 10% of the leopards' kills are lost to other predators, particularly lions, which is compensated for by similar levels of scavenging. Leopards minimize kleptoparasitism, which is a term describing parasitism by theft, by hoarding carcasses. Although caching behavior typically protects the carcass, 57% of tree cached carcasses in Kruger had scavengers in attendance, particularly spotted hyenas, whereas only 9% of carcasses dragged into thick vegetation in Kaudom attracted competitors. The records of giraffe calves being cached in trees can reflect the strength of leopards.

There are methods in studies that can minimize the biases towards larger prey with one of which being Jacob's index: $D = (r - p)/(r + p - 2rp)$. In Jacob's index, r stands for the proportion of the total kills at a region by a single species while p stands for the proportion of the abundance of that species of the total prey population. The resulting value ranges from -1 to +1 where -1 stands for complete avoidance where the carnivorous animal completely avoidance the prey while +1 stands for maximum preference where the prey animal stands as one of, if not the most, favored prey of the carnivorous animal. ±n where n is a complete number between -1 and +1 stands as the mean. Five animals that might be close to the preferred prey of the Waterbobbejan are the puku, water chevrotain, the red forest duikers, Thomson's gazelle, and the African civet. Using Jacob's index, the leopards' preference towards the puku is 0.98, the water chevrotain come in at 0.82±0.17, red forest duikers have a preference of 0.37±0.21, Thomson's gazelle are next with 0.14±0.21, and finally the preference of the African civet is -0.06±0.42. Using the same method, the five most proportionally common prey can be calculated, which are: Chital deer, with an astounding 64% of kills where they occur, impala with 48%, Thomson's gazelle with 33%, nyala with 19%, and springbok with 11%.

With only 20% of their most common prey items overlapping, there are drastic differences between the Waterbobbejan and leopards. However, diet is not the only evolutionary trait of leopards that shows that Waterbobbejan is far more than just a leopard. Unlike leopards, the cryptid isn't capable of climbing trees.

Climbing trees is a feature seen in both leopards and monkeys, although the latter seems to be better known for this capability. Samango monkeys are a perfect example for arboreal monkeys as they are the only exclusively arboreal monkeys that are native to South Africa.

Samango monkeys are restricted to a variety of forest habitats. Given the ongoing forest habitat loss and fragmentation, samango monkeys exist in isolated or semi-isolated forest fragments with a suspected low rate of dispersal. Although the estimated extent occurrence for samango monkeys is over 20,000 square kilometers with the area of occupancy being calculated as the amount of remaining natural habitat within forest patches greater than 1.5 square kilometers.

Samango monkeys utilize the canopy of evergreen forests, and their current distribution indicates very broad forest habitat tolerances. Being South Africa's only forest dwelling guenons, samango monkeys are associated with high-canopy evergreen forests. They inhabit several indigenous forest types, specifically afromontane forests, coastal forests, scarp forests, and riverine forests. Samango monkeys have also been observed in human-modified habitats, including pine plantations, residential gardens, and campsites. However, more research is required to confirm that samango monkeys are able to use modified landscapes to disperse between forest patches. They are able to utilize humans infrastructure to travel across their habitat, such as traveling along telephone and power lines and across roads. Nevertheless, samango monkeys seem to view human inhabited areas as more dangerous than their natural habitat, preferring to forage in indigenous forests if given experimental patches in both forests and gardens.

Samango monkeys are predominantly frugivores with 50 to 70% of their diet consisting of fruit. However, leaves or insects are their main source of protein. During the periods with a lack of fruits, samango monkeys would eat other plant parts such as flowers and buds. Samango monkeys have been recorded to eat exotic plant species that are either invasive or planted by people, consume human waste such as kitchen waste. Consequently, it is possible that samango monkeys may be considered as pests in some areas.

A key difference shown between the samango monkeys and the Waterbobbejan is that the Waterbobbejan does not require the same adaptations seen in samango monkeys. Because livestock are located in open regions, the Waterbobbejan is not restricted to forests as seen in samango monkeys. One of the main reasons that caused the evolution of the samango monkeys' arboreal lives is their diet, which is drastically different from that of the Waterbobbejan.

With key differences between diet, behavior, and physical appearances between the cryptid and both leopards and primates, there are several problems. Specifically, the cause of the death of the livestock was supposedly the Waterbobbejan. For that, the current status of South Africa's livestock industry must be analyzed.

Livestock is farmed throughout South Africa with the numbers and species varying according to the climatic conditions. The western 67% of South Africa is relatively dry and is primarily only suitable for livestock. Modern, highly advanced, and intensified systems of the livestock production are run parallel with the subsistence, communal pastoral systems.

Indigenous species and breeds that are well adapted to the environment have been preserved through many years in the livestock industry. Their adaptation and preservations is a unique benefit, and the breeds should be maintained in order to promote efficient extensive livestock production. The breeds' role is not merely regarded as breeding materials that are bound to be upgraded with exotic breeds, but are regarded to provide a broader and unique contribution to the national gene pool. Over several years, many people have imported breeds, types and their crossbreeds, that are now well adapted to the local conditions of South Africa, and have also made important contributions to the development of the livestock industry.

> There is a relatively higher number of sheep than cattle and goats with the exception of KwaZulu-Natal, Limpopo, Gauteng, and the North West. Even so, the sheep heavily outnumber the cattle and the

goats as shown in. It can be concluded by observations that sheep form 58% of the livestock in between these three animals. Cattle come in at 28% and the remaining 14% belong to the goats. This means that the sheep outnumber the cattle and the goats 2:1 and 4:1 respectively, outnumbering both animals combined at a ratio of 14:10.

This is an important factor as the fact that the livestock, especially these three animals that form the majority of the Waterbobbejan's diet, are decreasing. Between the years 1972 and 2002, the rate of change in the commercial production of livestock commodities n South Africa has been analyzed. It has been shown that the commercial productions with a decreased percentage are the mutton, which has decreased by 61%, wool, which decreased by 56%, butter, which decreased by 80%, condensed milk, which decreased by 52%, and skimmed milk powder, which decreased by 30%.

The data that has been analyzed is crucial to the theory of the Waterbobbejan since interviews with farmers in the 1860's concluded that the farmers lost mostly sheep, with cattle and goats not far behind. It has also been stated by the farmers that the livestock were killed in a method that was not even been possibly been attributed to the method of any known natural predators such as leopards, lions, hyenas or jackals. A particular incident states that a prize bull did not return home in the evening with the rest of the herd. When the farmer searched for the prize bull the next morning, he found the bull's remains completely dismembered. There are also stories of a huge ape-like creature, a strong animal that was sometimes seen leaping over enclosure walls with terrified animals under its arms. Even rarer however, is the fact that several other farmers experienced damage to their fruit trees that had been of all their oranges by a colossal ape-like creature according to eyewitnesses. Tswana farm laborers believe that the Waterbobbejan lives in caves, behind waterfalls, or near any body of water, stating that the cryptid would take its prey to a cave and devour it, ripping its prey to pieces.

If the claims of the Waterbobbejan preferring sheep as prey were true, then there would be a decrease in the amount of sheep compared to that of the cattle and goats. However, that is not the case as the sheep and animal ratio is 14:10. This can be thus concluded that the claims of the farmers are at least partially incorrect. It can also be stated that the method of hunting belonging to a Waterbobbejan is wrong. The dismemberment of the prized bull is mostly similar to that of humans, while the devouring and ripping the prey to pieces is similar to that of lions. Even so, there is currently no solid evidence to what the hunting method could be and if it is actually different from any known animal as there is too little evidence to exactly determine how the unknown animal killed the livestock.

Another reason why the hunting method could actually be that of a known animal is the fact that farmers are unreliable. Dr. Robert Brain, a famous paleontologist, of the Transvaal Museums located in Pretoria has investigated reports of the Waterbobbejan, and one incident supports the fact that farmers aren't a reliable source.

Dr. Robert Brain received a phone call from a farmer in the Mpumalanga Province who claimed to have shot, killed, and skinned a Waterbobbejan. When Dr. Brain visited the farmer, he disappointingly found that the trophy was of a samango monkey. The farmer had exaggerated the whole encounter to Dr. Brain on the phone, and contrary to Dr. Brain identification, the farmer continued to believe that he had indeed shot a Waterbobbejan. While this wouldn't necessarily apply to the rest of the farmers, but it should be noted that the farmers that had stated about the killings lived in the 1860's with far fewer knowledge about biology and forensic science than what is currently known today.

With a lack of reliability and solid evidence supporting the existence of the Waterbobbejan, the cryptid could be nothing more than a mere hoax.

CHAPTER V: THE EUROPEAN WILDMAN

As the name suggests, wildmen are supposed hominid cryptids living in Europe. While not as popular as its Himalayan and North American counterparts, wildmen are possibly even more likely to exist than the Yeti or Bigfoot due to one factor: location.

Throughout the evolution of primates, Europe is considered as a "hotspot" because there was a huge amount of primates that have evolved in the continent. The evolution of primates in Europe reached its peak in the Miocene epoch, and the natural history of Europe is an important factor to uncovering the identity of the European wildmen.

As part of the latter stage of the Cenozoic cooling, the Miocene epoch was a time of important climate and vegetation changes. During the Early Miocene, glaciation was uni-polar with an ice volume on Antarctica that is comparable to the present and a largely ice-free northern hemisphere. On the other hand, the first indications of northern hemispheric glaciation ultimately appeared during the Late Miocene, leading to the formation of the Greenland ice sheet in the Pliocene. A global intensification of orogenic movements considerably influenced the climate system. Likewise, the Late Miocene witnessed the development and spread of C_4-grasses, aridification of the interiors of continents, and the expansion of open landscapes. Although all of the events are considered to be linked, there has yet to be any proof of their casual interdependence.

Bruch et al. and various other authors have discussed the Miocene temperature patterns. Their data mainly demonstrate that the general cooling during the Miocene epoch brought a greater climatic differentiation, both spatially via increased seasonality of temperature. Additionally to the regional effects of paleogeography, temperature parameters reveal an increasing differentiation between the marine and continental climate conditions. However, this interpretation has largely been based on the examinations of temperature parameters. Miocene precipitation has been described so far only in terms of mean annual precipitation, whether through proxy-based reconstructions or in climate modeling.

Some of the climate reconstructions are based entirely on paleobotanical material. In total, 169 types of plants from the Miocene were selected in a climate reconstruction, including fourteen Burdigalian between 20.428 and 16.303 million years ago, forty one Langhian between 16.303 and 13.654 million years ago, forty Serravallian between 13.654 and 11.6 million years ago, thirty six Tortonian 11.6 to 7.251 million years ago, and thirty eight Messinian between 7.251 to 5.332 million years ago, excluding the Aquitanian stage due to the low amount of data. With the exception of the Burdigalian with only fourteen samples, the comparable numbers of floras represents all of the other stages. To increase the reliability of the results, the low-diversity floras were avoided.

The computer program ClimStat facilitates the application of the Coexistence Approach and the database Palaeoflora that contains NLRs of over three thousand Cenozoic plant taxa, together with their climatic requirements as derived from the meteorological stations that are located within the distribution areas of the taxa. According to the data available in the Palaeoflora database, the method allows the calculation of up to fifteen climate parameters.

Generally, the resolution and reliability of the resulting coexistence intervals increase with the number of taxa included in the analysis, and are relatively high in floras with at least ten taxa for which climate parameters are known. Because results of the Coexistence Approach analyses are intervals, the accuracy of calculated climate data correlates with the accuracy of the margins of the coexistence intervals. The accuracy of the coexistence intervals varies respectfully to the examined parameter. It is the highest for the temperature-related parameters where it is usually within the range of 1 to 2 degrees Centigrade and the mean annual precipitation of 100 to 200 millimeters. Other precipitation parameters are less accurate and mainly reflect overall trends.

In one particular study, the mean annual precipitation, monthly precipitation of the driest month, monthly precipitation of the wettest month, and the monthly precipitation of the warmest month have been calculated using the Coexistence Approach. Additionally, the mean annual range of temperature, which is the temperature difference between the warmest and coldest months, and the mean annual range of precipitation, which is the difference between the monthly precipitation of the wettest month and the monthly precipitation of the driest month, has been taken into consideration. Together, the given parameters provide detailed information on the precipitation patterns and a method of assessing continentality.

There are three major conclusions given by the study: the correlation of precipitation parameters, patterns of continentality, and the seasonality gradients.

Correlation coefficient and gradient of regression are two measures that reflect the relationships between two parameters. While the correlation coefficient expresses the reliability of the relationships of the data, the regressional gradients show the nature of the relationships.

The strongest correlation with the highest correlation coefficients in the entire data set arises between the monthly precipitation of the wettest and warmest months. Both parameters correlate with the other precipitation parameters to various degrees yet in a similar manner. When the different time intervals are compared, the correlation coefficients between the monthly precipitation of the wettest and warmest months are spotted to continuously increase from the Middle to Late Miocene Epoch. Thus, the monthly precipitation of the warmest month, as estimated using the Coexistence Approach, isn't just a good measure for the precipitation in summer, but it can be assumed to reflect the general trends and patterns found in precipitation parameters, especially for post-Langhian times.

It has been indicated that the gradient of regression between the monthly precipitation of the warmest month and other precipitation parameters is the steepest in the Burdigalian, Serravallian, and Messinian data sets. The low number of data available for the Burdigalian stage does not allow for further interpretation, but the generally increasing gradient found in the post-Langhian data indicates an increase in seasonal differentiation of precipitation towards a more noticeable precipitation peak in summer. On the other and, Langhian data show the lowest regressional gradient despite high absolute values for summer precipitation and a high correlation coefficient. This reflects generally humid conditions with no specific wetter or drier season in the summer.

Continentality is defined by low precipitation and strong seasonality of temperature. The history of continental climate in central Europe by showing the patterns of the temperature difference between the warmest and coldest months, as well as the mean annual precipitation, both as paleo-data alone and as differences of paleo-minus-modern data.

The mean annual range of temperature is lowest during Burdigalian and Langhian in Western Europe and increases beginning with the Middle Miocene. However, the coastal parts of the continent exhibit only minor changes, reflecting a persistent low seasonality of temperature influenced by the Atlantic Ocean. This pattern is the first evidence for a climatic differentiation in the Neogene epoch between oceanic and continental climates in Europe. Because of the buffering effect of the large Paratethys Sea, temperatures in eastern Europe also remain equable at least until the Tortonian. During the Miocene, patterns of the temperature difference between the warmest and coldest months were evidently not as strictly east west oriented as they currently are.

All of the Miocene precipitation data show typically humid conditions for the entire Miocene epoch. Apart from a general decrease in the mean annual precipitation, the Langhian and Tortonian appear to have wetter phases compared to the times before and after the Miocene epoch. Moreover, a progressive spatial differentiation appears in the late Middle Miocene epoch, in the Serravallian period, with relatively lower precipitation in the eastern section of central Europe.

Summer precipitation decreases beginning with the Middle Miocene, and all stages to some extent show a longitudinal differentiation with slightly lower values in Eastern Europe. This trend becomes clearer towards the Late Miocene and is the most important observed pattern change of all precipitation parameters that have been studied. Beyond changes in temperatures, shifts in the annual distribution of precipitation might have played a key role in post-Langhian climate changes.

The differences between modern and paleo-data document the very low degree of change in precipitation and the temperature difference between the warmest and coldest months during the Miocene epoch compared to the transformations that occurred post-Miocene and up to the present. Although the absolute values are obviously changing with the approach of the Messinian, the absolute values are still remote from those of today.

To aid in the analysis of the latitudinal gradients of precipitation and the difference between the warmest and coldest months as measures of continentality, fossil climate data are plotted against longitude. None of the data reveal statistically significant longitudinal gradients. Only the Messinian and Burdigalian data, alongside a small data sample, show a correlation between longitude and the monthly precipitation of the warmest month, and the Messinian and Serravallian data to some extent with the mean annual precipitation.

The lack of significant correlation between longitude and the temperature difference between the warmest and coldest months can mainly be traced to the buffering effect of the Paratethys, which suppressed any development towards temperature seasonality in the Pannonian Basin. Nevertheless, data from the Messinian stage show a slight increase in the temperature difference between the warmest and coldest months, together with the massive decrease in the mean annual temperature and the monthly precipitation of the warmest month from the western to the eastern points of all the data. With its combination of characteristics, the Messinian stands alone as the only stage to exhibit a tendency towards higher continentality in eastern Europe. All of the other Miocene stages show either no longitudinal changes, or not in the combination shown in the Messinian. In short, the evidence for continental climate in Eastern Europe first appears with the Messinian.

To date, the interpretation of the Miocene climate has been mainly based on the temperature parameters, with the Miocene rainfall generally described in terms of the mean annual precipitation. Overall, the data documents generally humid conditions, with all parameters exceeding the present values.

Langhian data show very high humidity with a peak in precipitation that does not match summer precipitation and therefore occurred sometime other than summer, and no longitudinal gradient. Meanwhile, Serravallian data show slightly less humid conditions with evidence of a summer peak in precipitation and no significant longitudinal gradient. Tortonian data evidence slightly higher humidity than the previous times with increasing evidence of a summer peak in precipitation, but no longitudinal gradients. Lastly, the Messinian data evidence less humid conditions with increasing evidence of a summer peak in precipitation, and first evidence of longitudinal gradients in the monthly precipitation of the warmest month and the mean annual precipitation. The succession testifies to a general decrease in precipitation since the middle Miocene and the onset of continental climate conditions in Eastern Europe in the late Miocene period.

Beyond the general decrease in the mean annual precipitation, the Langhian and Tortonian appear to have been wetter than the stages before and after them. Wetter conditions during the Langhian might be related to the Mid Miocene Climatic Optimum and are widely discussed in the literature. It is worth noting that the data show a strong annual range of precipitation in central Europe together with the highest value for all of the precipitation parameters, reflecting generally very humid conditions with some seasonality, except without a precipitation peak during summer.

The commonness of wetter conditions during the Tortonian than in preceding and successive stages has been less thoroughly discussed, mainly because the focus so far as been on comparing the conditions of the Middle and Late Miocene periods and on the general decrease in humidity. Only Böhme et al. describe the wetter conditions during the Tortonian than in the late Serravallian and early Messinian based on a study of herpetofauna and in good agreement with our results. On the other hand, their data indicate lower than the present mean annual precipitation during short-term dryer phases, where the data do not confirm such low values for any of the parameters analyzed. This might reflect the challenge of reconstructing very dry conditions on the basis of plant fossils. The lack of fossil floras preserved under dry conditions causes a strong bias in the data set with gaps towards southern Europe where plant proxy data are available only from moister regions of Spain and Italy. Moreover, faunas and floras usually come from different stratigraphic levels and taphonomic settings in central and Eastern Europe as well.

Miocene climatic changes after the Mid Miocene Climatic Optimum are testified in the data set by three major factors: increasing seasonality of temperature, the changes in the annual distribution of precipitation towards a precipitation peak in summer, and a late increase of longitudinal gradients of precipitation parameters.

The expansion of open landscapes during the Miocene in Europe has been the subject of widespread and sometimes heated debate based on the extinct organisms. Strömberg et al. provide a comprehensive review of the debate. Large mammal data suggest the presence of open environments in southern Europe since the early late Miocene. However, the data mainly confirm mosaic landscapes with open forests and don't support the notion of vast open landscapes. With the regard to plant fossils, only some data on phytoliths and pollen favor grass-dominated savannas or open woodlands are found. Showing that such habitats were widely established in the eastern parts of southern Europe during the Late Miocene based on phytolith analyses propose that relatively open habitats developed in Asia Minor beginning already in the Early Miocene. For Akgün et al., however, the analysis of Anatolian pollen descriptions documents the increasing abundance of open vegetation taxa exclusively during the Tortonian period. Additionally, Syabryaj et al. describe the vegetation development of the Ukrainian plain as a stepwise opening of the forests with the first steppe grasslands arising during the early Tortonian and expanding greatly during the late Messinian. Shatilova et al. observe the same pattern of development with an onset of opening of vegetation at the Khersonian stage for a shallow-marine succession in eastern Georgia. Only Kovar-Eder et al. provide a broader view of the European vegetation history based on a quantitive analysis of an immense floristic data set from Europe. The authors summarize that for the Tortonian as "first records of xeric grasslands are found along the northern margin of the Black Sea", while during the Messinian "open woodlands increasingly appeared in central and southern parts of Italy and Greece."

The patterns correspond with the observations made of the earliest data to support continental climate in Eastern Europe during the Messinian. However, the data is not able to correlate with an onset of opening of vegetation during or prior to the Tortonian period. Either that development in vegetation is not related to the climate parameters analyzed, or, conceivably other perhaps non-climatic parameters played a role, as discussed controversially by various authors. It is possible that the implications of certain taxa as evidence for both dryness and openness are overestimated by the qualitative approaches or underestimated by the method due to the usually wide climatic ranges of the taxa. Hence, the data might lack the required resolution and/or spatial coverage to completely decode the influence of continentality on vegetation and to correlate climatic and vegetation data statistically. This could apply mainly on the southern and eastern Europe with their still low data coverage, but also points back to the methodological question of the ability of plant proxies to detect dry events.

Nevertheless, climatic data that can quantify continentality are a useful basis for describing the expansion of grassland in Eurasia. Further studies on a wider geographic scale, with an increased focus on southern Europe and central Eurasia, will assist in quantifying the development of precipitation patterns and understanding their influence on the history of landscape opening during the Neogene epoch.

The conditions caused primates to flourish into Europe, and many species of primates and early hominids to appear in Europe.

GIANT APES: THE CULTURALLY DIVERSE CRYPTIDS
SECTION 1: MIOCENE APES

The Miocene epoch, which lasted 18 million years between 23 and 5.3 million years ago, marks the evolution of primitive apes. The main paleographic feature that is relevant to primate origins involved cycles of expansion and reduction in the size of primate habitats in the Mediterranean and Eurasia. Land bridges formed and disappeared between Africa, Europe, and Eurasia. Because of the cooling temperatures, tropical rain forests transitioned to a mosaic of woodland savannah with trees and grass, and savannas, which only have grass. The cooling temperature also resulted in the expansion of savannah-woodland environments in Africa and Eurasia. During this time, old world monkeys and apes diverged, and the apes underwent an adaptive radiation into eighty to a hundred species thereafter.

There are several differences between the Miocene apes and modern apes however, and the main morphological differences being in the placement of the scapula, the depth of the ribcage, the extension of the elbow joint, the shoulder joint, the spine, the hip joint, the arm to leg ratio, and the hands. Firstly, the placement of the Miocene apes' scapula is on the side while the scapula of modern apes is located on the back. Secondly, Miocene apes have a deep ribcage depth while modern apes have a shallow ribcage depth. Thirdly, the Miocene apes have a partial extension in their elbow joint, while the modern apes have a full extension in their elbow joint. Fourthly, while the shoulder joint of Miocene apes is restricted, the shoulder joint of modern apes is mobile. Fifthly, the spine of Miocene apes is long and flexible, while the spine is short and rigid. Sixthly, the hip joint of the Miocene apes is restricted while the hip joint of modern apes is mobile. Seventhly, the arm to leg ratio is found to be equal in the Miocene apes while modern apes' arms are longer than their legs. Finally, the hands of the Miocene apes are small while modern apes have large hands.

During the early Miocene 23 to 16 million years ago, ape-like primates evolved in eastern Africa. Out of the many species that have been discovered in the early Miocene epoch, the most abundant specimens are those in the genus Proconsul. These diurnal apes weighed in between 17 and 50 kilograms, marking a major increase in body size within the primate clade. Despite the increase in body size, estimates of the brain size of the Proconsulidae family do not differ from those seen in Miocene monkeys. Proconsul had sexually dimorphic canines and a mostly frugivorous diet, meaning that it mostly fed on fruits. Most Proconsulidae were quadruped, although some species were more arboreal than others. Overall, Proconsul exhibits a mixture of both ape and monkey features. For example, the Miocene primates have a relatively large body size, thick molar enamel, and no tail, which are attributed to apes. However, the Proconsul species still retain some features of monkeys, such as certain morphological aspects of their backbone and pelvis. Other ape-like species from the early Miocene such as Afropithecus and Kenyapithecus also show a variety of features of apes, monkeys, and completely unique features. Thus, the early Miocene apes were likely to be better suited to traveling on top tree branches rather than hanging or swinging below tree limbs as seen in modern apes. Nonetheless, the cranial and postcranial features of the early Miocene apes provide strong evidence for transitional changes in the primate evolution.

During the middle Miocene 16 to 11.6 million years ago, African apes moved across land bridges to colonize Eurasia and eventually parts of Eastern Europe and Asia. Furthermore, David Begun hypothesized in 2003 that some European apes migrated back to Africa during the early stages of the middle Miocene. However, the African immigrants did not last long as most species went extinct by 13 million years ago. Some of the best-known apes from the middle Miocene are Dryopithecus from Europe and Sivapithecus from modern day India and Pakistan. Dryopithecus and Sivapithecus were both large, with body mass estimates between 20 and 90 kilograms. Just as in modern apes, Dryopithecus and Sivapithecus had teeth suited for masticating fruit pulp, shortened snouts, and long, strongly built jaws. A reduced snout represents a derived primate feature that reflects a reduced reliance on olfaction. The cranium of Sivapithecus is very similar to that of modern orangutans, although the postcrania of the two species have a few in common. Dryopithecus brains were similar in size and proportions to those seen in modern chimpanzees, some of the most intelligent animals on the planet. Morphological interpretations of Dryopithecus and Sivapithecus postcrania reveal strong relationships to the suspensory locomotion utilized by modern apes. To conclude, Dryopithecus and Sivapithecus were more similar to apes than to monkeys.

The late Miocene epoch lasted for 6.3 million years from 11.6 to 5.3

million years ago, and experienced a continual gradual decline in the global temperatures. In many regions throughout the globe, temperate and tropical forests reduced in size due to the lessening of rainfall and global temperatures. Although these changes likely resulted in the extinction of many ape species in Europe and parts of Asia, other species migrated to or continued to exist in the tropical zones of Africa and Southeast Asia. Consequently, some extinct apes from the middle Miocene survived into the late Miocene such as Dryopithecus and Sivapithecus. Additionally, paleoanthropologists have recovered remains of some latecomers to the ape lineage such as Oreopithecus, Ouranopithecus, Lufengpithecus, and Ankarapithecus. Oreopithecus might be the best known of all Miocene apes due to the plentiful fossils discovered in Italy. Oreopithecus had the classic catarrhine dental formula of 2.1.2.3, and its dental morphology indicates a folivorous diet, a rather remarkable adaptation in the ape lineage. Most modern apes, with the exception of some species of gorillas, are more frugivorous than folivorous. Although the body mass of Oreopithecus is estimated to be approximately 30 kilograms, it had a small brain, which is a trait not shared by modern apes. Examination of the several postcranial remains of Oreopithecus indicates suspensory locomotion similarly to modern apes.

Ouranopithecus was a large ape that weighed 70 to 110 kilograms from what is now known as Greece. Ouranopithecus likely had a diet consisting of hard, gritty foods. Like Sivapithecus, the cranium of Ouranopithecus is similar to that of modern orangutans. Currently, there are no postcranial remains of the ape. Another large ape from the late Miocene is Lufengpithecus. Lufengpithecus weighed in around 50 kilograms and lived in what is now known as southern China. This species had extreme sexual dimorphism in their dentition and cranial features, similarly to Dryopithecus. Lastly, Ankarapithecus was an 82-kilogram ape from modern day Turkey. This species had teeth with very thick enamel and large jaws that it probably used to eat hard or gritty foods. Ankarapithecus retained the primitive trait of a prognathic cranium, which differs greatly from the flatter, more derived cranium of Sivapithecus.

With the evolution of apes and the migration towards Europe via land bridges, apes thrived during the epoch. Because the Miocene epoch ended only 5.3 million years ago, there are several theories that circulate around the evolution of apes during this time and turning into the European wildmen.

SECTION 2: THEORIES

There are several theories surrounding the European wildmen, the most prominent of which being the Neanderthal theory.

Neanderthals lived during the Ice Age, taking shelter from the ice, snow and the otherwise unpleasant weather in Eurasia's plentiful limestone caves. Many fossils of Neanderthals have been discovered in caves, therefore leading to the popular view of the Neanderthals as "cave men". This early species of humans had a short, stocky stature as an evolutionary adaptation for cold weather, since their stature consolidated heat. According to the Smithsonian Institution, the Neanderthals' wide nose helped to humidify and warm the cold air, though this matter is still being debated. The American Museum of Natural History states that other differences between Neanderthals and humans are a flaring, funnel-shaped chest, a flaring pelvis, and robust fingers and toes. Neanderthals' brains, however, grew at a similar rate to those of humans, and were about the same size, if not larger. It has been estimated that approximately 1% of Neanderthals had red hair, light skin, and perhaps even freckles.

For a long period of time, scientists have theorized that Neanderthals grew up at a faster rate than humans, reaching maturity sooner and dying younger than humans. In 2008, however, Proceedings of the National Academy of Sciences published evidence that humans and Neanderthals mature at the same rate.

While the Neanderthals do match the wildmen in many aspects, could the Neanderthals really still exist? To know the possibilities of that matter, the history of the early modern human colonization and its effects on Neanderthals must be studied.

After a span of over 200,000 years of the Neanderthals' successful adaptation to the glacial climates of northwestern Eurasia, they abruptly disappeared between 30,000 and 40,000 years ago. Around the time that the

Neanderthals disappeared, populations that are all but identical to modern humans replaced them. Recent research suggest that the roots of the dramatic population replacement can be traced back to events occurring in another continent, with the distinctively modern human remain and artifacts in eastern and southern Africa.

Studies of both the mitochondrial and the Y-chromosome DNA patterns in the populations in the modern world point to the genetic origins of all present-day populations within one limited area of Africa around 150,000 years ago, followed by their scatter to other regions throughout the globe between 60,000 and 40,000 years ago. The results are further reinforced by the discoveries in the early 2000's that reveal the skeletal remains of modern populations in different regions. Discoveries at Herto, Ethiopia, were reported in 2003, confirming the presence of early forms of modern humans in Africa around 160,000 years ago. Meanwhile, the earliest discoveries of distinctively modern populations in both Europe and most parts of Asia can be dated to no earlier than 40,000 to 45,000 years ago.

In Europe, the most dramatic support for the patterns came from the recovery of a number of relatively well preserved sequences of mitochondrial DNA from several skeletal finds of Neanderthals and early humans. Analyses of seven Neanderthal specimens supplied pieces of mitochondrial DNA that are drastically different from those of all known modern populations in either Europe or any other region throughout the world. The segments are equally as different from those recovered from five specimens of early modern populations found in European sites.

It can therefore be clearly concluded that there was very little, if any, interbreeding between the Neanderthals and the intrusive early humans. If interbreeding did occur, all of the genetic traces of the interbreeding between the two species and were thereafter eliminated from the European gene pool. The mitochondrial DNA evidence recovered from the Neanderthal specimens further suggests that the initial evolutionary separation between the Neanderthals and the populations, which would eventually evolve to modern humans, goes back to around 300,000 years ago. The fossil evidence from Africa and Europe also supports this conclusion.

The general assumption made in the past has been that the separation from Africa to the other regions throughout the globe is represented by the widespread distribution of 'Aurignacian' technologies, tracing back continuously from the neighboring areas of the Near East, made up of Israel, Lebanon, Syria, and other countries, through most areas of eastern and central Europe, to the Atlantic coasts of Spain and France, which all broadly fall within the time range of 40,000 to 35,000 years ago in conventional radiocarbon terms.

Significantly, the Aurignacian period shows a sudden 'boom' of the modern cultural behavior. These features include the first complex and carefully shaped bone, antler, and ivory tools, an unexpected increase of perforated animal teeth, far-travelled marine shells, carefully shaped stone and ivory beads and other forms of personal ornaments, and outstandingly diverse forms of both abstract and figurative art, which range from the engraved outlines of animals and the representation of the male and female sex organs to the ivory statuettes of animal and human figures from southern Germany and the cave paintings found in the Chauvat cave in southeastern France.

Collectively, the Aurignacian period reflects an explosion in the clearly emblematic behavior among the Aurignacian populations in Europe and western Asia that the Neanderthal communities of the Middle Paleolithic period in Europe lack. Symbolic communication and expression shown in that complexity would require complex language and therefore, a brain built similarly, if not identically, to our own.

However, there is a massive difficulty of identifying considerable and anatomically distinct human skeletal remains that are associated with the Aurignacian technologies. Several skeletal remains generally in an anatomically modern form which were ascribed at first to the Aurignacian have recently been proven on the foundation of radiocarbon dating of the bones to depict intrusive funerals into the Aurignacian levels from much later levels, particularly those from Velika Pećina, Croatia and the remarkably early Aurignacian site of Vogelherd located in southern Germany.

Despite the removal of the skeletal remains from Vogelherd and Velika Pećina, there is a variation of more fragmentary skeletal remains from at least five to six well-documented contents in Europe and western Asia which distinctively point to the existence of diagnostically modern populations that fell within the same time range as the Aurignacian occupation and are directly associated with Aurignacian archaeological matter in several instances. Currently, the best-dated finds are the remains of three generally modern individuals that were reported from the Pestera cu Oase Cave, Romania, and the remains of a complete juvenile skeleton unearthed from the levels directly underlying the long Aurignacian sequence at Ksar Akil, Lebanon. The juvenile skeleton was dated on the basis of both archaeological evidence and excessive radiocarbon measurements to at least

40,000 years ago. In Western Europe, a fragmentary maxilla was discovered from the Kents Cavern, Devonshire and was directly dated to 30,900 ± 900 years ago. Two characteristically modern mandibles were unearthed in Les Rois, France, and were closely associated with the early Aurignacian levels on the Les Rois site dating to around 32,000 to 35,000 years ago.

Churchill and Smith suggested that the incomplete mandible and other remains from the initial pre-Aurignacian levels at Palaeolithic Kira, Bulgaria were probably of anatomically modern form, with radiocarbon data ranging from 39,000 to 43,000 years ago. Lastly, the pair of distinctively anatomically modern crania and other skeletal remains from the Mladeč site, Czech Republic, have been dated on the foundation of the radiocarbon measurements of the associated calcite deposits to around 34,000 to 35,000 years ago. This means that the skeletal remains are almost certainly associated with a range of generally Aurignacian bone artifacts.

Archaeological research has shown evidence for an seemingly double pattern of colonization made by early modern populations across Europe along two separate routes.

The first route is depicted by 'classic' Aurignacian technologies that have been examined. In other words, the first route is represented by Aurignac itself and marked by a range of distinguishable tool forms, including typical nosed and carinated scraper forms, heavily edge-trimmed Aurignacian blades, and the highly distinctive split-base bone and antler spear-head forms. These technologies were distributed across a broad arc that crosses western, central, and southwestern Europe and extend into the directly adjoining areas of the Near East. Although the available radiocarbon dates for the technologies show a broadly similar pattern across the region, centering around 34,000 to 38,000 years ago, there are are strong indications that the origins of the technology can be identified to be much earlier at sites in southeastern Europe and the eastern Mediterranean region than anywhere in central and western Europe with both areas extending back to at least 40,000 years ago.

The second route of dispersal is distributed mainly along the Mediterranean coast of Europe, extending from at least northeastern Italy to the Atlantic coast of northern Spain. Although these industries are often referred to as archaic or proto-Aurignacian in literature, the industries show an absolutely different pattern of technology from that of Aurignacians. The technology of the industries is mostly dominated by small, carefully shaped bladelets that might have served as the tips and barbs of composite spearheads or arrowheads. These industries represent a sharp break with the immediately preceding Neanderthal technologies in the regions. The most persuasive origins for the industries' technologies lean towards those occurring in the sites located in the Near East, dating back to around 38,000 to 40,000 years ago. Both these industries and those of the original Aurignacian period seem to be fundamentally obtained from the preceding Ahmarian and Emiran technologies of the Near East, which are represented by the unusually lengthy early Palaeolithic sequence at Ksar Akil around 45,000 to 47,000 years ago. At Ksar Akil, the initial Upper Palaeolithic levels are associated with the burial of a generally anatomically modern skeleton. Thus, the levels might indicate the earliest appearance of modern populations in the region shortly after their deduced dispersal from Africa shortly beforehand.

It is implied that contact, and therefore potential interaction, between Neanderthals and modern humans is inevitable. One point that seems clear, however, is that the appearance of a number of apparently modern features of technology among some of the last Neanderthal communities of central and western Europe can be shown to closely correspond with the appearance of early Aurignacian populations in nearby regions of central Europe, and probably with those along the Mediterranean coast.

Such patterns of behavioral interaction and technological transfer between the Neanderthals and modern populations are precisely what would be predicted by the basis of examples of recent ethnic contact circumstances, without regard of the respective cultural and cognitive capacities of the two populations. If the evolutionary trajectories of the Neanderthal and modern populations had been separated from at least 300,000 years as all available genetic and anatomical evidence suggests, then the possibility of some divergence in the neurological structures over the period should be ruled out. Identically, the possibility of some small degree of interbreeding between the Neanderthals and modern humans cannot be excluded on the basis of either the current anatomical or DNA evidence, making it appear to be plausible in anthropological and demographic terms.

All traces of the distinctively Neanderthals patterns of mitochondrial DNA and anatomical features have disappeared relatively rapidly from Europe populations, however. This might reflect a straightforward case of direct competition for territorial and resource conflicts between Neanderthals and modern humans. The modern humans' demonstrably more complex technologies and organization gave them a strong competition advantage over the Neanderthals. Some of the rapid climatic wavering that have been documented spanning in this range might have also played a critical role in the demographically competitive circumstance. That, together with the extinction of many of the Neanderthals' prey animals, must have caused their downfall, with the possibility of the Neanderthals still existing today being next to impossible.

Now excluding the Neanderthals as the European wildmen, the wildmen would either be Oreopithecus or Dryopithecus. A short calculation however, marks down another creature. Using relative growth rate of marine animals that grew b 150-fold in 542 million years, Oreopithecus would weigh in at 133 kilograms and standing in at 233 centimeters, or roughly 7'8". On the other hand, Dryopithecus would weigh in at a mere 79 kilograms and stand in at 90 centimeters. This would exclude Dryopithecus as the European wildmen because it would be merely too small if it was still alive today.

Oreopithecus, also known as the 'mountain ape', has been referred to as the 'abominable coalman' because the skeletal remains of the Oreopithecus were found in the brown coal deposits of Tuscany in northern Italy roughly 14 million years ago. Another reason to the nickname is because of its humanlike features. This animal had a monkey's snout and ankle bones, yet had the brow ridges similar to those of an ape. Oreopithecus had a flat and small face with conical canines and patterns on the molar teeth were similar to those of hominids. The mixture of monkey and human like characteristics might be explained by the consideration of oreopithecines as an independent lineage.

The classification of the oreopithecids, however, is problematic. Some paleontologists regard them as Old World monkeys related to Mesopithecus, but others point to their attributes similar to those of humans and apes, such as their probable ability to brachiate, the ability to use their arms to swing from branch to branch, and even walk upright. They are nearly unquestionably an evolutionary blind alley, though, whose advanced features are a result of convergence.

Further details about the Oreopithecus can be inferred, however. Since its remains were preserved in beds of soft brown coal, Oreopithecus likely lived in forested, riverside swamps. It is likely that it survived on a diet of leaves, shoots, and fruits of a diverse amount of plants. Oreopithecus might have been arboreal because its arms were longer than the legs. The spine and hip bones suggest that Oreopithecus was also capable of walking, or at least lope along in an upright stance.

As previously stated, the Late Miocene Epoch had faced a gradual decline in the global temperature and a decrease in rainfall, hence causing the contraction of the global area of temperate and tropical forests. This most likely affected the forested swamps where Oreopithecus lived in. If this was the case, then it would be nearly impossible for Oreopithecus to still be alive in Europe. Further evidence that this would cause Oreopithecus to go extinct, habitat loss must be studied.

Habitat loss can generally be placed into three categories: Habitat destruction, fragmentation, and degradation. Habitat destruction is the complete removal of trees and plants, instantaneously changing the landscape. A prime example of habitat destruction is mass deforestation by cutting down trees. Habitat fragmentation is the type of habitat loss that takes by place by altering the land in a manner that confuses the animals and disrupts their natural way of living.

Fragmentation occurs when roads are created and attractions are placed in the midst of woodlands and other natural areas. By fragmenting habitats, areas might not be completely destroyed but the fragmentation still causes environmental chaos. Fragmentation can separate animals from others and their food sources, occurring both in the water and on the land. Beneath the water, structures such as dams tend to isolate species from each other, making it more difficult for the species to mate and find food. For the many animals that migrate, the fragmentation of habitats removes this advantage.

Finally, habitat degradation is a form of destruction that occurs by pollution that causes habitats to be destroyed because pollution changes the quality of air, water, and the land, while becoming a breeding ground for toxins. Degradation of an environment causes invasive species to appear, which naturally contribute to the downfall of the natural animals and plants. While this occurs, native species begin to go extinct from the negative environmental changes.

These types of habitat loss have several effects. First, habitat destruction is the leading cause of extinction for numerous species, and is the reason that many modern animals are endangered. Secondly, the process of habitat loss often happens so rapidly that there is not enough time for the animals to adapt to such drastic changes. Thirdly, as wildlife is displaced, it changes the way that animals live and behave. Young animals need to be cared for in multiple ways in the wild as they have to be protected from predators. Fourthly, the eventual result is death and the inability to reproduce offspring that would carry on the species. Fifthly, many plants would no longer grow because the composition and quality of soils is instantaneously changed, taking away the nutrients and space that plants require to grow. Finally, the current status of habitat loss is hurting not only animals and plants, but humans as well. This is because habitat is contributing to global warming and climate change.

Out of the mentioned effects of modern habitat loss, five of those match the extinction in the Late Miocene epoch, with the exception of which being global warming. Yet, according to the effects stated, there should be a possibility where Oreopithecus had managed to adapt to the changing climate. For that, however, the equation $sqrt((4(r-d)DyDx(a^2))/Dy + Dx(a^2))$ must be solved and the solution must be compared to other primates. For the sake of the supposed anatomy of the wildmen, the solution will be compared to that of real data used on Neanderthals, perhaps the closest thing to the European wildmen.

Using the equation, the rate that Oreopithecus would have to go to evolve to the wildmen is 460% the rate of the Neanderthals. This further proves that it is impossible for Oreopithecus to be the European wildmen.

With scientific and mathematical proof that the European wildmen are neither Neanderthals, Dryopithecus, nor Oreopithecus, there is no possible manner to think that the European wildmen are a species that has not yet been confirmed. Even with both scientific and mathematical evidence, the last piece of evidence actually comes from festivals.

Across Europe, festivals span from the beginning of December until Easter, and correspond to the Christian holidays. However, the rituals themselves in the festivals would often predate Christianity. The festivals supposedly rekindle the connection between Europe and the nature's 'rhythms'. These 'rhythms' are that monsters loom in the shadowy heart of the winter, but so does the promise of spring's rebirth and fertile crops.

The roots of the festivals are difficult to trace. People would supposedly don costumes that hide their face and 'conceal their true forms'. Afterwards, the people would take to the streets where their disguises allow them to cross the line between human and animal, real and spiritual, and death and rebirth. António Carneiro, who dressed as a devilish careto for Carnival in Podence, Portugal stated to National Geographic that a man would assume a dual personality and become something mysterious.

Photographer Charles Fréger set out to capture what he calls 'tribal Europe' over a span of two winters across nineteen countries. The forms of the costumes that Fréger chronicled varied, even between villages. In Corlata, Romania, men would dress as stags reenacting a hunt with dancers. Meanwhile in Sardinia, Italy, goats, deer, boars, or even bears might play the sacrificial role.

This festival is the core of the European wildmen, with a version appearing throughout Europe. However, Gerald Creed, who has studies mask traditions in Bulgaria, has stated that all people know that they shouldn't believe that costumes and rituals have the power to 'banish evil and end winter'. This would mean that the wild men are a complete hoax, and both the scientific and mathematical evidence support Gerald Creed's statement.

It would also be easily concluded by Gerald Creed's statement, the festivals, and the mathematical and scientific evidence that the wildmen are nothing but a mere hoax made by pure imagination.

CHAPTER VI: THE AUSTRALIAN YOWIE

The Australian Yowie isn't as popular as any of the previously mentioned versions of Bigfoot, with the exception of the African Waterbobbejan. Historical accounts of the Yowie refer to two types of the cryptid in Australia.

The first species is said to grow between 1.8 to 3 meters, or 6 to 10 feet, tall and weigh up to 454 kilograms, or 1,000 pounds. This species is described to resemble a huge ape-like man with talons instead of fingers. Compared to the North American Bigfoot, the Yowie is believed to have a face and head closer to that of a primate, as well as being capable of walking upright. It has also been as described as being more aggressive and dangerous towards humans compared to its North American counterpart. The second species is described to be smaller, with a height between 1.2 and 1.5 meters, or 4 to 5 feet, tall. Some people believe that the Yowie is an ancient species of hominids that has avoided extinction. In local cave art, the hominids have been depicted as tall, hairy figures beside smaller Aboriginal figures. In order to know the possibility of the theory, the basic background and geological history of Australia must be studied.

Australia is an island continent located between the Indian and Pacific oceans. With an area of 7,686,850 square kilometers, Australia is the world's sixth largest country. It has a diverse landscape, which includes tropical rainforests, deserts, snowcapped mountains, agricultural land, and admirable beaches. Some of Australia's best-known natural features are Uluru, also known as the Ayers Rock, and the Great Barrier Reef. Most Australians are the descendants of immigrants from several European countries such as Britain and Ireland. By 2013, native Aborigines make up a mere 2% of the population. This translates to roughly 452,000 native Aborigines out of the 22.6 million citizens in 2013.

The long geological history of Australia began in the early Cambrian epoch 540 million years ago. The early Cambrian epoch experienced major changes, and to understand how major the differences are, a comparison between the early Cambrian epoch and the Vendian epoch 620 million years ago must be made.

During the Vendian epoch 620 million years ago, the Earth was extremely unusual compared to the Earth seen in modern maps. Land covered much of what is now known as the Pacific Ocean, while seas covered the region where Europe, Asia, and Africa are now located. In the Vendian epoch, two main continents predominated the Earth, which are not so creatively known as Northern and Southern Gondwana. Northern Gondwana is made up of what are now known as India, Antarctica, and Australia. Meanwhile, Southern Gondwana consists what are now known as Africa, the Americas, and parts of Asia. Currently tropical regions, such as West Africa and parts of South America, were congregated during the Vendian epoch near the South Pole and were excessively glaciated.

By 540 million years ago in the early Cambrian epoch, the drastic differences between the Vendian and the Cambrian epoch. Both of the major continents of the Vendian epoch, which have temporarily combined to momentarily form a supercontinent called Pannotia, had broken up. During the early Cambrian epoch, a remainder of Pannotia called Gondwana stretched nearly from pole to pole. Gondwana was made up of what are now known as China, India, Australia, Antarctica, Africa, and South America. Two important landmasses that aren't components of Gondwana were Laurentia, which includes the majority of North America, and Siberia. A growing mid-oceanic ridge between the two islands and Gondwana thrusted them on a long journey northwards.

Half a billion years ago in the Late Cambrian epoch, the northern hemisphere was nearly completely empty, with the exception of a submerged remainder of modern Russia near the North Pole. Near the South Pole, Avalania, consisting of parts of Britain, Ireland, and the eastern American seaboard, Iberia, consisting of Portugal and Spain, and Armorica, consisting of other remains of Western Europe, were located underwater off the coast of Gondwana about 13,000 kilometers, or 8,000 miles, away from their current position. Eastern Australia was located on the northern coast of Gondwana. Eastern Australia included a series of mountain belts that formed as the ancient core of the continent collided with thin silvers of continental shelf called 'micro continents' starting about 500 million years ago.

460 million years ago near the end of the Ordovician epoch, most continents began to grow in land area, with intense volcanic activity adding land to the east coast of Australia, as well as to parts of Antarctica and South America. During this time, parts of Gondwana began to become detached. The lasting parts moved towards the south, where North Africa laid directly over the South Pole.

In the Silurian epoch 420 million years ago, most of the continents laid in the southern hemisphere. Gondwana, which consisted of what are now known as South America, Africa, Australia, and India, was located near the South Pole. Avalania, a continental fragment that consisted of much of the eastern seaboard of America, closed up an ocean known as the Iapetus Ocean as it approached Laurentia, which is consisted of most of what is now known as North America. At the south of Avalania, the Rheic Ocean began to open up. In eastern Australia, substantial volcanic eruptions occurred from the mid-Silurian to the Devonian epoch.

360 million years ago in the Devonian epoch, two massive supercontinents were slowly drifting towards each other. In the south was Gondwana, consisting of what are now known Australia, Antarctica, India, Africa, and South America. In the north is Laurentia, which consists of what are now known as North America and northern Europe. Shallow seas flooded the area now known as the American Midwest, while Iberia, which consists of what are now known as Spain and Portugal, is an island off Laurentia's south coast. In the Devonian epoch, mountains continued to build up in Australia. The southeastern coast of the continent collided with a chain of volcanic islands, forming a chain of mountains. New rivers flowing from the mountains carried sediment to basins in the center of the continent, forming distinctive Devonian rocks. Shallow seas covered parts of western Australia. Over time, the seafloor sediments have been compressed to form the characteristic reef limestone and mudstone. Within these rocks, fossils of extremely diverse communities of fish and other marine life.

The Carboniferous epoch began 354 million years ago. For the first time throughout the geological history, the supercontinent Pangaea can be perceived. During this time, Pangaea has started to form from the Laurentia, which consists of North America and Europe, with Gondwana. Before the collision, Gondwana had rotate clockwise, so that the eastern part of Gondwana, consisting of India, Australia, and Antarctica, moved southward while its western part, consisting of South America and Africa, moved northward. The rotation of Gondwana opened up a new ocean called the Tethys Ocean in the east, and closed up the Rheic Ocean in the west. At the same time, the ocean in between Baltica and Siberia was closing, paving the way for another collision of continents. In the early Carboniferous epoch, seas that were warm enough for coral reefs to grow lapped the north of Australia. The coral reefs are currently visible as limestone. Dense forests in the eastern part of the continent laid down thin layers of coal. In the later Carboniferous epoch, Gondwana moved south and the climate cooled, with glacial deposits in southern Australia being evidence for the movement. In a span of 30 million years, Australia moved south by twenty degrees of latitude.

The Permian period lasted from 290 to 248 million years ago. During this span of time, Pangaea had finally assembled when the ancient island continent of Siberia joined up with the rest of the major landmasses. Thus, the crustal plates stretched from pole to pole. Just as Pangaea had finished forming, fragments began to break off the southern continent of Gondwana. An active ocean ridge opened up, creating new ocean floor. As a newly formed sea widened during the Permian period, parts of the edges of Gondwana were pushed way from Pangaea. The micro continents that formed and traveled northwards in this manner included places such as what are now known as Tibet and Malaysia. The end of the Permian period was marked by the biggest mass extinction in history, killing off 90% of all species. A sharp drop in sea levels and huge lava flows in Siberia means that much of the species on land and on shallow shelf seas have been impacted, with most going extinct. It has been estimated that less than 5% of marine animals and 33% of terrestrial animals have survived. The huge lava flows in Siberia would have also speeded up climate change.

240 million years ago in the Triassic period, a vast ocean dominated one hemisphere while Pangaea dominated the other hemisphere. Pangaea consisted of what are now known as North America, Europe, North Asia, Africa, South America, India, Australia, and Antarctica. During the Triassic period, Pangaea gradually moved northward. Since the end of the Permian period, sea levels had risen again and therefore allowing a slow recovery of marine life, such corals. On the land, tropical coal-forming forests and swamps diminished as climates gradually became warmer and drier.

In the Jurassic period 170 million years ago, the Earth was a warmer, less varied place than it currently is. It is likely that there were no ice caps at the poles for much of the period, and the mild conditions made for much higher sea levels, resulting in a smaller area of land but extensive shallow continental seas, which teemed with life. Pangaea had started to split up, and familiar modern landmasses such as North America and Eurasia started to appear. The North and South Atlantic Oceans both began to open up, while the Tethys Ocean began to close.

The Cretaceous period spanned almost 80 million years between 142 and 65.5 million years ago. During this time, the Earth began to take on a familiar look as the land that had once made up Pangaea pulled apart. The newly formed Atlantic Ocean had extended north and southward, separating Africa and Eurasia from the Americas. Meanwhile, Africa, India, Antarctica, and Australia began to drift apart about 120 million years ago as new oceans formed between the components of Gondwana. Asia was still unfamiliar, with land from its southern edge such as India and Indochina still separate islands.

In the later Cretaceous period around 90 million years ago, the Atlantic Ocean separated the New World, which consists of the Americas, from the Old World, which consists of Europe, Asia, and Africa. Much of Asia was assembled with the exception of India. During this time, India was still attached to Madagascar. Meanwhile, Australia still languished in the deep south while attached to Antarctica. The formation of the Southeast Indian Ridge between Antarctica and Australia began the opening of an ocean between the two continents. Australia was slowly pushed northward as new ocean floor was created on spread on either side of the ridge. After the Cretaceous period, Australia continued to separate from the other major landmasses, creating a drastic difference in the hominid theory.

SECTION 1: HOMINIDS

The key features of human evolution, also known as 'hominization', on which attention has focused, include both the physical and cultural developments. As hominization has proceeded, differences in anatomy have been less significant than changes in life of way, usage of the environment, and interpersonal relations. Physical developments have included changes in the locomotion and posture of hominids, most notably the upright stance as clearly seen in humans, a bipedal locomotion, legs longer than the arms, and reduced big toes. The growth of the pelvis and the birth canal in order to adjust to larger-brained babies, increased manual dexterity due to the lengthened thumbs and the ability to hold small objects delicately between the thumb and forefinger, and modifications of the head, are all physical developments in hominization.

Compared with the more primitive primates, hominids have smaller jaws and teeth, with their teeth being more thickly enameled, and the faces of hominids are flatter than those of other primates. Hominids have lost the ridges of bone located above the eyes and the crest on top of the skull, but the brains of hominids are relatively larger compared to the rest of their body. The brains of hominids have also been developed to be more complex than the brains of other primates.

Cultural development of hominids includes the formation of groups, cooperative work, tool making, the harnessing of fire, the making of sculpture and painting, and burial rites. Each development should be observed as existing in a complicated feedback relationship. For example, the hands disused for locomotion allowed a greater facility in the production of tools, which supports increased hand-eye coordination and the development of the brain. Tool usage also puts a value greatly on improvements in child-rearing, social organization, and communication. A long-term social group that lengthens infant care and supports its members throughout adulthood is more capable of acquiring, sharing, and accumulating experience. Social groups that cooperate intimately are also likely to use tools more efficiently, and to improve their design and manufacture.

Many crucial cultural developments, such as language and social structure, do not fossilize. Therefore, paleontologists can only guess about the cultural developments based on the activities, which leave traces, such as the evidence of burial rites. Because there is such a selective record to go on, great caution must be exercised in interpretation. Generally, a lot is known about stone tools because of their capability to fossilize, but tools made of animal and plant materials, such as leather and plaited-fiber bags have little information left for the paleontologists to interpret.

To further know about the possibility of the Yowie being a hominid, the origins of mammals, and hence primates, must be put into place. Mammals evolved from cynodonts roughly 220 million years ago. Therefore, mammals were already an ancient lineage by the time Australia separated from the rest of the continents, let alone the time that the dinosaurs went extinct roughly 65.5 million years ago in the Cretaceous-Tertiary extinction. This marked the end of the 'Age of the Dinosaurs' called the Mesozoic and the start of the 'Age of the Mammals' called the Cenozoic.

Hominids appeared in the Miocene epoch 6 to 7 million years ago with the appearance of Sahelanthropus tchadensis in Chad, Central Africa. A nearly complete cranium knows Sahelanthropus nicknamed Tourmai as well as a number of fragmentary lower jaws and teeth. It is currently unknown whether Sahelanthropus is bipedal or quadruped. This species had many primitive apelike features such as a small brain size with an area of approximately 350 cubic centimeters, along with many other features such as the brow ridges and small canine teeth. The mixture of apelike and humanlike features, along with the fact that it comes from roughly the same time when the hominids are believed to have diverged from chimpanzees, suggests that Sahelanthropus is close to the common ancestor of humans and chimpanzees.

Orrorin tugenensis was discovered in western Kenya. Fossils of this species include fragmentary arm and thighbones, lower jaws, and teeth that date back to 6 million years ago. The limb bones of Orrorin were 150% larger than Lucy, a famous specimen of Australopithecus, and suggest that it was roughly the same size as a female chimpanzee. It is claimed by the discoverers of this species that Orrorin was a human ancestor adapted to both bipedalism and tree climbing. Given the fragmentary natures of the Orrorin remains, other scientists have been skeptical of these claims so far, mainly from the 2001 paper written by Aiello and Collard. A later paper made by Galik et al. in 2004 has found further evidence of bipedalism in the femur of Orrorin.

In September of 1994, Australopithecus ramidus was named from fragmentary fossils that dated back to 4.4 million years ago. More complete cranium and partial skeleton were discovered in late 1994 and the species was reallocated to the genus Ardipithecus based on that fossil. This fossil was extremely fragile, causing the excavation, restoration, and analysis of the fossil to take fifteen years. It was published in October of 2009, given the nickname 'Ardi'. Using Ardi, Ardipithecus has been estimated be 120 centimeters, or 3'11", tall, and weighed about 50 kilograms, or 110 pounds. The skull and brain of the animal were quite small, with a size comparable to a chimpanzee. Although Ardipithecus was bipedal on the ground, it was not as well adapted to bipedalism as the australopithecines. This might be because Ardipithecus were quadruped in the trees. It lived in a woodland environment with patches of forest, indicating that bipedalism might not have originated in a savannah environment.

Between 1997 and 2001, a number of fragmentary fossils were discovered that date back to 5.2 to 5.8 million years old, and were originally assigned to a new species called Ardipithecus ramidus kadabba. However, they were later assigned to a new species called Ardipithecus kadabba in 2004. One of these fossils discovered is a toe bone that belongs to a bipedal creature, but analysis of the toe bone shows that it is a few hundred thousand years younger than the rest of the fossils, and so the identification of the toe bone with Ar. kadabba is not as firm as the other fossils discovered.

Australopithecus anamensis was named in August 1995, and consists of 9 fossils that were mostly found in 1994 from Kanapoi, Kenya. A dozen fossils, mostly teeth, were found in 1988 from Allia Bay, Kenya. A. anamensis lived 4.2 to 3.9 million years ago, and has primitive features in the skull, yet it has advanced features in the body. The teeth and jaws of A. anamensis are very similar to those of geologically older apes. A partial tibia is a strong evidence of bipedalism, and a lower humerus is extremely humanlike. Although the skull and skeletal bones are believed to be from the same species, it is yet to be confirmed.

Australopithecus afarensis is a species of Australopithecus that existed between 3.9 and 3 million years ago. A. afarensis had an apelike face with a low forehead, a bony ridge over the eyes, a flat nose, and no chin. They had protruding jaws with immense back teeth. The cranial capacity of A. afarensis varies between 375 and 550 cubic centimeters. A. afarensis' cranium is similar to that of a chimpanzee with the exception for more humanlike teeth. The canine teeth are much smaller than those of modern apes, but are larger and more pointed than those of humans, and the shape of the jaw is a mixture of that of apes and that of humans. However, the pelvis and leg bones of A. afarensis resemble more closely to those of modern humans. This makes A. afarensis unmistakingly bipedal, although they were more adapted to walking rather than running. Their bones show that they were physically strong. Australopithecus afarensis show sexual dimorphism with females being substantially smaller than males. The height of A. afarensis varied between approximately 107 centimeters, or 3'6", and 152 centimeters, or 5'. The finger and toe bones of A. afarensis are curved and proportionally longer than those of humans. However, the hands of A. afarensis are similar to humans in most of the other details. Most scientists consider this as evidence that A. afarensis was still partially adapted to climbing in trees while others consider it as evolutionary baggage.

Australopithecus africanus existed 3 to 2 million years. Similar to A. afarensis, A. africanus was bipedal with the latter having a slightly bigger body size. The brain size of A. africanus might have also been slightly larger, ranging anywhere between 420 and 500 cubic centimeters. This is slightly larger than chimp brains, despite a similar body size, but is still not advanced in the regions necessary for speech. The back teeth of A. africanus were also slightly bigger than those of A. afarensis. Although the teeth and jaws of A. africanus are much larger than those of humans, they are far more similar to the teeth of human than those of apes. The shape of A. africanus' jaw is fully parabolic similarly to humans, and the size of canine teeth is further reduced when compared to the canine teeth of A. afarensis.

Australopithecus garhi was named in April 1999 from a partial skull. The skull of A. garhi differs from the previous species in the combination of its features, notably the extremely immense size of its teeth, especially the rear teeth, and primitive skull morphology. Some nearby skeletal remains might belong to A. garhi, showing a humanlike ratio of the humerus and femur; bat an apelike ratio of the lower and upper arm.

Australopithecus sediba was discovered at the site of Malapa, South Africa, in 2008. Two partial skeletons were found with one belonging to young male while the other belongs to an adult female. The two partial skeletons date back to 1.78 to 1.95 million years ago. It is claimed by the discoverers of A. sediba to be transitional between A. africanus and the genus Homo, and because it is more similar to Homo than any other australopithecine, it is a possible candidate for the ancestor of the genus

Homo. A. sediba was bipedal with long arms suitable for climbing, but had a number of humanlike features in the cranium, teeth, and pelvis. The young male's skull has volume of 420 cubic centimeters, and both fossils are short with a height of about 130 centimeters, or 4'3".

All of the previously stated species of Australopithecus are known as gracile australopithecines because their skulls and teeth are not as large and strong as the incoming species, which are known as the robust australopithecines. Despite this, they were still more robust than modern humans.

Australopithecus aethiopicus existed between 2.6 and 2.3 million years ago. A. aethiopicus is known from the Black Skull discovered by Alan Walker, and a few other minor specimens that may belong to the same species. This species might be an ancestor of A. robustus and A. boisei, but it has a puzzling mixture of primitive and advanced traits. The brain size of A. aethiopicus is very small at 410 cubic centimeters, and parts of the skull, particularly the hind portions, are very primitive, most resembling A. afarensis. Other characteristics, such as a big face, jaws, and single tooth found, as well as the largest sagittal crest, a bony ridge on top of the skull to which the chewing muscles attach, in any known hominid, are more similar to A. boisei.

Australopithecus robustus had a body similar to that of A. africanus, but had a larger and more robust cranium. A. robustus had lived between 2 and 1.5 million years ago. The large face is flat or dished, with no forehead and large brow ridges. It had relatively small front teeth, but had big grinding teeth in a large lower jaw. Most specimens show sagittal crests. Thus, its diet would have been mostly coarse, tough food that required a lot of chewing. The average brain size of this species is about 530 cubic centimeters. Bones that have been excavated with the skeletons of A. robustus indicate that they might have been used as digging tools.

Australopithecus boisei lived between 2.1 and 1.1 million years ago. This species was similar to A. robustus, but the face and cheek teeth of A. boisei were even larger with some molars reaching up to 2 centimeters across. The brain size of A. boisei is similar to that of A. robustus at around 530 cubic centimeters. This caused a few experts to consider that A. boisei and A. robustus to be variations of the same species.

With many species belonging to the Australopithecus genus, it is indeed very impressive and holds very strong candidates for the Yowie. However, the genus Homo is even more impressive and holds even more details that would supposedly point to the Yowie of Australia.

Homo habilis, which translates to 'handy man', was called so because of the evidence of tools discovered with the remains of the species that lived between 2.4 and 1.5 million years ago. This species was very similar to australopithecines in several ways. The cranium of H. habilis is still primitive, but it projects less than A. africanus. The back teeth of H. habilis are small, but are still considerably larger than those of modern humans. It is estimated that the average brain size of H. habilis is considerably larger than that of australopithecines at 650 cubic centimeters. Brain size varies between 500 and 800 cubic centimeters, which overlaps the australopithecines at the low end and H. erectus at the high end. The brain shape of H. habilis is also more humanlike. Broca's area's bulge, which is essential for speech, is visible in a brain cast of H. habilis, indicating that it is likely that H. habilis is capable of rudimentary speech. H. habilis is thought to have been around 127 centimeters, or 5' tall, and about 45 kilograms, or 100 pounds, in weight. It is possible that females have been smaller than males.

Homo georgicus is a species that was named in 2002 to contain fossils that were found in Dmanisi, Georgia, that seem to be intermediate between H. habilis and H. erectus. Three partial skulls and three lower jaws were discovered, all of which date back to about 1.8 million years ago. It is estimated that the brain sizes of the skulls vary between 600 and 780 cubic centimeters. As estimated from a foot bone, the height of H. georgicus might have been roughly 1.5 meters, or 4'11".

Homo erectus existed between 1.8 million and 300,000 years ago. Similarly to H. habilis, the face of H. erectus has protruding jaws with large molars, no chin, thick brow ridges, and a long, low skull. The brain size of H. erectus varies between 750 and 1,225 cubic centimeters. Early specimens of H. erectus average about 900 cubic centimeters, while later ones average in 1,100 cubic centimeters. The skeleton is more robust than that of modern humans, implying that H. erectus had greater strength. Body proportions of this species vary. For example, the Turkana Boy is tall and slender, similarly to modern humans in the same region, while the few limb bones found of Peking man indicate a shorter, sturdier build than the Turkana Boy. Study of the Turkana Boy skeleton indicates that H. erectus might have been more efficient at walking than modern humans, whose skeletons required to adapt to allow for the birth of larger-brained infants. H. habilis and all australopithecines are restricted on Africa, while H. erectus has been discovered in Africa, Asia, and Europe. There is evidence that H. erectus might have used fire, and their stone tools are more advanced than those of

H. habilis.

Homo antecessor was named in 1977 from fossils found at Atapuerca, Spain that date back to at least 780,000 years ago. This would make H. antecessor the oldest confirmed European hominids. The mid-facial area of antecessor is advanced, while other parts of the skull such as the teeth, forehead, and brow ridges are much more primitive. Many scientists doubt the validity of H. antecessor partially because a juvenile specimen defines it.

Archaic forms of H. sapiens first started to appear about five hundred thousand years ago. The term covers a diverse group of skulls that have features of both H. erectus and modern humans. The brain size is larger than H. erectus and smaller than most modern humans, averaging in at about 1,200 cubic centimeters, and the skull is more rounded than the skull of H. erectus. H. sapiens had a skeleton and teeth that are generally less robust than H. erectus, but are more robust than modern humans. Many still have large brow ridges and receding foreheads and chins. There is no clear dividing line between the late H. erectus and archaic H. sapiens, and many fossils that were dated back to 500,000 to 200,000 years ago are difficult to classify as which species.

One species of H. sapiens is the Neanderthal. Neanderthals lived between 230,000 and 30,000 years ago, and had an average brain size slightly larger than that of modern humans. However, this probably correlated with their greater bulk. The brain case of Neanderthals is longer and lower than that of modern humans, with a marked bulge at the back of the skull. Like H. erectus, Neanderthals had protruding jaws and receding foreheads. The chin of Neanderthals was usually weak. The mid-facial area of Neanderthals also protrudes, which is a feature that is not found in H. erectus or H. sapiens, and may have been an adaptation to the cold. There are other minor anatomical differences between Neanderthals and modern humans, with the most unusual of which being some peculiarities of the shoulder blade and the pubic bone. Neanderthals mostly lived in cold climates, and their body proportions are similar to those of modern cold-adapted people, being short and solid with short limbs. Their bones are thick and heavy, and show signs of powerful muscle attachments. A large number of tools and weapons belonging to Neanderthals were discovered, being more advanced than those of H. erectus. Western European Neanderthals usually have a more robust form, and are sometimes called the 'classic Neanderthals'.

Homo floresiensis was discovered at Flores, an Indonesian island, in 2003, from several individuals. The most complete fossil is of an adult female with a height of about 1 meter, or 3'3", and a brain size of 417 cubic centimeters. Other fossils indicate that the female had an average size for the species. It is thought that H. floresiensis is a dwarf form of H. erectus as it is common for dwarf forms of large mammals to evolve on islands. H. floresiensis was completely bipedal, used stone tools and fire, and hunted dwarf elephants which are also found on the island.

Modern forms of H. sapiens first began to appear about 195,000 years ago. It is estimated that the average brain size of modern humans is 1350 cubic centimeters. The forehead rose sharply, the eyebrow ridges are very small and are more usually absent, the chin is prominent, and the skeleton is very gracile. About 40 thousand years ago, with the appearance of the Cro-Magnon culture, tool kits started to become much more advanced, using a more diverse array of raw materials such as bones and antler, and containing new implements for making clothing, engraving, and sculpting. Fine artwork in the form of decorated tools, beads, ivory carvings of humans and animals, clay figurines, musical instruments, and spectacular cave paintings started to appear over the upcoming twenty thousand years.

Within the last hundred thousand years, the long-term trends towards smaller molars and decreased robustness can be discerned. The face, jaw, and teeth of Mesolithic humans that lived about ten thousand years ago, are about 10% more robust than ours. Upper Paleolithic humans roughly thirty thousand years ago, are around 20 to 30% more robust than modern humans in Europe and Asia. These are considered as modern humans, although they are considered as primitive. Interestingly, some modern humans, the aboriginal Australians to be exact, have tooth sizes more typical of archaic H. sapiens. The smallest tooth sizes re found in those regions where food-processing techniques have been used for the longest time. This is a probable example of natural selection that has occurred within the last ten thousand years.

With many species of the genus Homo and Australopithecus, there are a lot of possibilities to what the Yowie could possibly be. However, the same major flaw that gives the European wildmen an advantage should be placed within the hominid theory: location.

SECTION 2: AUSTRALIAN HOMINIDS

It is estimated that while hominids appeared 6 to 7 million years ago, hominids only appeared in Australia 65 thousand years ago. A team of archaeologists from the University of Queensland came to the conclusion by excavating a rock shelter in Majedbebe, northern Australia, during digs that were conducted in 2012 and 2015. Among the artifacts found in the region were stone tools and hatchets, which indicates an advanced understanding of weapon making. The study's authors claimed that similar hatches did not appear in other cultures for another 20 thousand years.

Instead of using radiocarbon dating, which is accurate for dating material up to 45 thousand years ago, researchers used a technique called optically stimulated luminescence, or OSL for short. The technique is applied to mineral grains and determines when it was last exposed to light, thus indicating how long an artifact has been buried. Initially, the artifacts dated back to only ten thousand years ago. As the archaeological team dug further into the shelter, they found tools that date back to 30, 40, and 65 thousand years ago.

In order to reach Australia, Aboriginal people required undertaking roughly 97 kilometers, or 60 miles, from the surrounding regions. It is also stated that it is possible that early Australians walked towards the northern regions from Papua New Guinea when sea levels were significantly lower than it is today.

Chris Stringer, the author of *The Origin of Our Species* and researcher at the Natural History Museum in London, noted that pushing back the time period for the human migration indicates that the early humans may have not been in direct conflict with other hominids and animals that have been previously thought. Earlier studies suggested that the arrival of humans corresponded with the extinction of several species. As previously stated, human migration has also been attributed to the decline of Neanderthals.

A possible candidate for the Yowie is Homo floresiensis. H. floresiensis, nicknamed the Hobbit, lived in Indonesia about 100 to 50 thousand years ago. Specimens of the Hobbit reach at 1.07 meters, or 3'6". Even after being generous to the species and assuming that it would grow at the same rate as marine animals that grew by 150-fold in 542 million years, H. floresiensis would grow by a mere 2%, making it 1.1 meters or 3'7" in modern times. Meanwhile, the Australian Yowie has a minimum height of 1.2 meters, or 4', which is five inches taller than H. floresiensis. One equation that has previously been mentioned can be given to show the plausibility of H. floresiensis growing in size to be the size of the Yowie, which is the rate of evolutionary growth: $sqrt((4(r-d)DyDx(a^2))/Dy + Dx(a^2))$. After solving for H. floresiensis, then it would require at least 7.8, nearly 8 times higher than the Neanderthal. Even taking in the factor that Neanderthals would have the same mutation rate, the Hobbits would have a rate 2.2 times higher than that of Neanderthals as estimated in the previous chapter.

Another more plausible species to be the Australian Yowie is Homo erectus. H. erectus lived in Africa, and Western to Eastern Asia 1.89 million to 143 thousand years ago. This species is the oldest known early human to have possessed proportions similar to modern humans, with relatively elongated legs and shorter arm compared to the size of the torso. The proportions are considered as adaptations to terrestrial life, indicating the loss of the earlier arboreal adaptations, with the ability to walk and possibly even run long distances. Compared to earlier fossils, there is an expanded brain case relative to the overall size of the cranium. The most complete specimen of H. erectus is the previously mentioned as the 'Turkana Boy', which is a well-preserved skeleton that dates back to around 1.6 million years ago. Microscopic study of the Turkana Boy's teeth indicates that he grew up at a similar growth rate as a great ape. The appearance of H. erectus in the fossil record is often associated with the earliest hand axes, which are the first major innovation in the stone tool technology.

Two major details that support H. erectus being an Australian hominid that evolved to the Yowie are the fact that they were available during the opening of the land ridge and that H. erectus is associated with the first major innovation in technology. However, this would cause H. antecessor, H. neanderthalensis, and H. sapiens to appear in the theory as Australian hominids.

Although the lineage of the genus Homo is currently heavily debated, there is one possibility that would make the hominid theory truly bizarre.

Theoretically speaking, H. antecessor, a close relative to H. erectus, is the ancestor of both H. sapiens and H. neanderthalensis. This would mean that the Yowie appeared after the divergence between the two species. Therefore, it would only seem to be possible if the Yowie appeared after the divergence of the Neanderthals and modern humans. When looking at Australia, it is clear that the Yowie would evolved to be a type of Neanderthal, which reveals a significant connection to the European wildmen.

While the European wildmen are nothing more than mere fiction, the connection that the Neanderthals are capable of evolving to the European wildmen and migrating to Australia shows that there is indeed some connection between the European and Australian cultures. This is indeed the case when looking at history as more than 95% of Australians originated from Europe.

Nonetheless, if this were the case, then the Neanderthals would have migrated to Australia in a rate of at least four hundred meters per year, a feat that is completely plausible. However, as the theories would state, there are a few problems with the three top candidates of the Yowie.

SECTION 3: THEORIES

There are three major theories about the Yowie: the Neanderthals, early modern humans, and H. antecessor. Because these form what seems like a giant jigsaw puzzle with evidence supporting both sides, the first type of evidence to be considered is the mathematical evidence.

Firstly, there is size. Using the method of relative growth rate of marine animal that grew by 150-fold in 542 million years, the current size of the three species would be estimated. Neanderthals would increase in height from 160 to 162.9 centimeters, or 5'3" to 5'4". Early modern humans would grow from 180 to 183.2 centimeters, or 5'11" to 6'. Finally, H. antecessor would grow from 170 to 207.4 centimeters, or 5'7" to 6'10", ever since it had 'supposedly' gone extinct 780 thousand years ago. However, none of the primates, whether alive or extinct, had faced this growth rate, and the growth rate of the hominid lineage is more accurate. Using this method, Neanderthals and early modern humans would grow by a mere 0.56 centimeters, or 0.22 inches, while H. antecessor grows by a mere 0.65 centimeters, or 0.26 inches. This would mean that the first method makes the evolutionary growth rate of the hominids 5.2 to 57.5 times higher than the actual rate of the second method. Even so, they both face problems since all three species would still be larger than the 152.4 centimeters, or 5 feet, limit of the Australian Yowie.

Another important factor to consider is the fossil evidence. While early modern humans did indeed evolve to modern humans, the fossil evidence would support the fact that the Neanderthals and H. antecessor have gone extinct. Even if taking the assumption that the fossils have become rarer, there is still a high chance that at least one bone from the upcoming five thousand generations of Neanderthals and the thirty nine thousand generations of H. antecessor would be discovered. There is also the fact that they would pass through Europe, Asia, and into Australia if the migration theory is correct. If so, then the human colonization problem would once again appear as it has been discussed about the European Neanderthals. While being extremely generous to the two species and consuming the fact that there was no competition between them and modern humans, there would sill be a type of interaction known as mutualism. Mutualism is a type of symbiotic relationship, a relationship between the individuals of separate species in an ecosystem, in which both species would benefit.

In the mutualism between early modern humans and both Neanderthals and H. antecessor, the relationship would most likely end up in the trade of weapons. This would therefore cause more interbreeding than the current evidence shows, and there would be more diversity in tool usage. Taking in the worst-case scenario that there would be competition, then there would be mass graves of one of the species that would show signs of attack from the tools of another, as well as marking the end of a specific type of tools and artifacts. In either way, there is a massive lack of evidence that would otherwise be clearly discovered.

The last factor to be discussed that solidifies the fact that Neanderthals and H. antecessor went extinct is climate. Neanderthals have lived in cold environments during the Ice Age, which had a mean temperature of -11 degrees Centigrade. This is much lower than the average temperature of 30 degrees Centigrade in modern Australia. These significant changes would most likely cause Neanderthals to go extinct. If the Neanderthals did not go extinct by the change, then there would be new species to be discovered that would be slightly more adapted to warm climates than Neanderthals. On the other hand, H. antecessor would have faced drastic cycles that would simply be too harsh for a mammal its size.

An expert on the Yowie called Rex Gilroy has claimed to investigate over 3,000 cases and believes that the Yowie is related to the North American Bigfoot. In the 1970's, Gilroy's research on the Yowie for several newspaper and magazines introduced the topic of the cryptid to the public. Despite the numerous sightings and eyewitness accounts, some researches have come to the conclusion that evidence for the Yowie is so rare that the cryptic is most likely a hoax, the same results the two main points have been leaning towards. In 2006, authors Tony Healy and Paul Cropper published a book about the cryptid called, '*The Yowie: In Search of Australia's Bigfoot*'. Healy and Cropper admitted that there is little evidence to support the existence of such a creature, which is once again the same conclusion reached by the second main point about the Yowie.

It can be thus concluded that the Yowie is a hoax as the probability of the

Yowie being a mere hoax is much higher than the possibility of the cryptid being an actual animal.

CHAPTER VII: THE CONCLUSIVE THEORY

There has been six versions of Bigfoot that have been discussed: The North American Bigfoot, the Himalayan Yeti, the South American Mapinguari, the African Waterbobbejan, the European Wildmen, and the Australian Yowie. Four of these, which are the Yeti, the Mapinguari, the Wildmen, and the Yowie, are a mere hoax with unknown origins. On the other hand, Bigfoot and the Waterbobbejan are misidentified animals.

Although the scientific, mathematical, and historical evidence all support the main theories applied on each version, there is still a piece of evidence that should be argued: witnesses. While there are many branches to the explanation for the mass of eyewitnesses, the main focus is on the expectation and repetition.

The expectation effect is an effect that is produced by that knowledge about what would occur afterwards, and is sometimes called the 'top-down control'. Meanwhile, the repetition effect is an effect that is produced by the sharing of some property between the current and preceding stimuli, and is sometimes called the 'bottom-up priming'. Logically, these two effects are orthogonal, but they are confounded. In a study made by Liqiang Huang and Harold Pashler in 2005, expectation and repetition refer specifically to the expectation and repetition of the feature values of the target and distractors on the defining features.

In previous work made on the repetition priming as it affects the singleton search, Maljkovic and Nakayama had subjects search for a singleton that could be defined in one of two ways in 1994. For example, a white target has black distractors, and a black target has white distractors.

Maljkovic and Nakayama discovered that even while the repetition of the feature value was as likely as the alternation, it nonetheless sped the responses of the subjects. Maljkovic and Nakayama reported that the priming effect was only induced by repetition and hardly affected by expectation. However, their study focused solely on priming effects in response time to sustained displays rather than on the accuracy of perception of brief displays. Thus, whether the effect arose from changes in perceptual or post-perceptual processing stages still remains undetermined.

To examine the effect, Huang and Pashler focused on work about the effects observed for response accuracy with brief displays. In their final experiment, they compared accuracy effects with findings observed for response time with unlimited viewing time. As has been frequently noted, response time indexes both perceptual and post-perceptual stages, whereas accuracy using very brief displays measures only the perceptual processing stages. In their study, the goal was to uncover the effects within the perceptual stage. Thus most of Huang and Pashler's experiments served to assess subjects' accuracy of perception of brief displays.

Prinzmetal, McCool, and Park provide a recent example of the method in which the response time and accuracy measurements differ in 2003, which examined the Stroop Effect. The Stroop Effect occurs with the naming of colors. Generally, when the subjects are required to respond to the colors of words that are the names of different colors such as 'Red' where the words' meanings notably interfere with the naming of their colors. The naming is substantially slower than it is in a neutral condition in which the words aren't related to the colors. However when Prinzmetal et al. performed the same experiment with brief displays to test the accuracy of color naming, they found a negligible difference between colored words and non-colored words. This result suggests that the interference found in the Stroop effect is post-perceptual.

In the experiments made by Huang and Pashler, 50 undergraduates from the University of California, San Diego, received credit in a psychology course for their participation in the project.

The stimuli were presented on a 1,024 X 768 MAG DX-15T color monitor driven by an Intel Pentium IV 1.8-G computer. The subjects viewed the displays from a distance of about 60 centimeters and entered responses using the keyboard. The program was written in Microsoft Visual Basic 6.0 and run on the second version of Microsoft Windows 98 using timing routines that were tested using a digital timer. Each search display had twenty lines. Each line was 1.25 degrees long and 0.21 degrees wide. All lines were white with a luminance greater than thirty cd per meter square against a black background with a luminance less than 0.2 cd per meter square. All of the lines with the exception of one had a single particular orientation on the screen, which are the distractors, and the remaining line, which is the target, had a different orientation. Ten lines were randomly placed in each of two regions that are 6.65 degrees wide and 14.36 degrees high. The regions were located on both the left and right halves of the display, with each spaced 2.13 degrees from the center of the screen. The location of the target, whether it was on the left or right half of the display, determined the correct response.

Each trial began with a small green fixation cross-presented in the center for 400 milliseconds. After a black interval of 400 milliseconds, the display appeared. The subjects were instructed to fixate on the cross and subsequently to search for the target in each given display. In the first four experiments, each of the displays was masked after short stimulus duration. On the other hand, the fifth experiment, the displays remained until the subjects responded. The subjects decided whether the target was in the left or right half, and responded by either pressing the 'j' or the 'k' button, with the 'j' button being the left half, and the 'k' being the right. A tone sounded for about 500 milliseconds to indicate whether the response was correct, and the next trial began 400 milliseconds later. Each subject performed ten blocks out of a hundred trials each, with the first 2 blocks being excluded as practice. Different block conditions alternated for each of the subjects, and were counterbalanced among them.

In the first experiment, the task was to find and report the location of a vertical target line among the horizontal distractors or vice versa. There were two alternating types of block. In a homogenous block, the feature setting, whether it was in a horizontal or vertical orientation, of the target and distractors were the same in every trial. Therefore, there might be all vertical distractors and all horizontal targets in a homogeneous block, or vice versa. In a random block, the feature settings were selected randomly for each trial. The feature settings of homogenous blocks alternated and were counterbalanced across subjects.

The results show that the accuracy was higher in homogenous blocks than in random blocks but there was no obvious difference between repetitions and alternated target-featured trials. It is indicated by the results that the repetition and expectation of feature settings are jointly sufficient to produce a significant in perceptual advantage. However, repetition by itself produces very little effect. Either there are two plausible interpretations of the result: Either expectation has a significant effect, whereas repetition has none, or an effect occurs when repetition and expectation are simultaneously present.

The second experiment was intended to arrange the conditions so that the subjects would have an expectation for a target feature value. However, the target feature value would not be repeated. As in the first experiment, the subjects' task was to find and state the location of a vertical target among horizontal distractors or vice versa. There were two alternating types of block as in the first experiments. First is an alternation block, where the feature settings in each trial were reserved from those of the previous trial. This type of 'predictable alternation; has been used previously by Maljkovic and Nakayama in 1994 to measure the role of expectation. In a random block, the feature settings were decided randomly for each trial.

Accuracy in alternated blocks was shown to be around equal to the accuracy in random blocks. In random blocks, there was no clear difference between repeated and alternated trials. Of the two possible interpretations of the first experiment, the second experiment evidently favors the latter, where a significant perceptual advantage is gained only by the joint operation of both the expectation and repetition. For this reason, the priming effect is hereafter called the expectation-repetition effect. An across-experiments analysis of variance, or ANOVA for short, indicated a significant interaction between expectation and repetition, where $[F(1,18) = 9.79, p < 0.1]$. The absence of any repetition effect in the random blocks of the experiment confirmed the similar result of the first experiment.

In the third experiment, the task was to locate and report the location of a diagonal target tilted at 45 degrees to the left, among either vertical or horizontal distractors. In a homogenous block, the distractors' orientation was constant in each trial. In a random block, the distractors' orientation as determined randomly for each trial. In both types of block, the distractors' orientation was constant. In the fourth experiment, a significant advantage of homogenous blocks over random blocks, as the target feature values were predictably in homogeneous blocks and varied randomly in random blocks.

Both experiments indicate that homogenous blocks had no significant advantages in accuracy over the random blocks. In random blocks, there was no clear difference between repeated trials and alternated trials. The results of the third experiment suggest that there is no distractor-feature inhibition effect. The results of the fourth experiment, meanwhile, indicate only a small target-feature facilitation effect. However, the latter approaches significance and should therefore be acknowledged as a potentially genuine effect. Nevertheless, it is considerably smaller than the expectation-repetition effect identified i the first experiment [interaction in across-experiments ANOVA: $F(1,18) = 6.12$, $p < 0.05$]. The expectation-repetition effect can't be explained by either the distractor-feature inhibition or the target-feature facilitation, or by their algebraic summation. Evidently, the effect relies on some interaction between processing of the distractor and the target features.

Finally, the purpose of the fifth experiment was to determine whether he critical difference between the first experiment and the research of Maljkovic and Nakayama in 1994, was the use of brief displays with accuracy measures. The fifth experiment was identical in method to the first experiment, except that the displays remained present and unmasked until subjects responded, and the response time was measured instead of accuracy.

In the fifth experiment, the mean response times were calculated. For homogeneous blocks, the response time was 480 milliseconds with an error of 2.7%. For repeated trials in random blocks, the response time was 548 milliseconds with an error rate of 1.8%. For alternated trials in random blocks, the mean response time was 594 milliseconds with an error rate of 2.8%. Responses in homogeneous blocks were significantly faster than those in random blocks with an effect of 91 milliseconds [$F(1,9)= 51.23$, $p < 0.001$]. In random blocks, repeated trials were significantly faster than alternated trials with an effect of 46 milliseconds [$F(1,9)= 9.62$, $p < 0.02$].

There is an important difference between the first and fifth experiment. Repetition in random blocks produced no clear effect on accuracy in the first experiment, whereas it produced a significant effect on response times in the fifth experiment. It seeks unlikely that the substantial 46 millisecond effect in the fifth experiment would reflect an equivalent difference in perceptual processing, since the effect on accuracy was a mere 0.87% in the first experiment. Besides the fact that the accuracy in the first experiment reached 71% at the stimulus duration of 127 milliseconds excludes the possibility that accuracy generally increases very slowly over time. The accuracy of 71% should mean that accuracy was near its maximum level of sensitivity. The repetition effect on speed reflects a different underlying cause from the perceptual expectation-repetition effect of the first experiment. The former presumably arises exclusively in some post perceptual whereas the latter arises in perceptual processing operations.

Reanalyzing the data from random blocks in the first two experiments, Huang and Pashler plotted the repetition effect as a function of several successive repetitions of the target feature value. Huang and Pashler have found no significant accumulation. They did not analyze runs of more than four successive repetitions since that type of trial was incredibly rare and did not allow a reliable estimation of the repetition effect. It is safe to state that the accumulation of successive repetitions cannot explain the expectation-repetition effect shown in the first experiment. Therefore, expectation must be a component of their explanation.

A tentative account is proposed that relies on the notion of what is termed as a feature divider in featured singleton detection. The feature divider is a mechanism that implements a categorization rule that divides orientation future space into two parts. As a result of the feature divider's operations, elements with some set of orientations are 'highlighted', whereas the remainder is ignored. In each of Huang and Pashler's trials, the feature divider had to locate a parameter that could successfully one item while it ignored the others. In the first experiment, the feature divider could use the same parameter for all trials of a homogeneous block. During random blocks, however, it had to switch parameters back and forth. In the third and fourth experiments, a single parameter could be used for all trials of a random block, a parameter that distinguished both vertical and horizontal lines from 45 degrees left-tilted lines, so there was a bit of advantage in homogeneous blocks.

It is believed that the experiments' different stimulus durations reflect the relative difficulty of the tasks. The difference in duration is not the cause, but rather the result of the expectation-repetition effect. The advantage was always available, even in the random blocks. This conclusion is consistent with the feature divider account.

If the repetition was an attribute of the feature divider that it preserved in the parameter of each trial just past, then the accidental repetition found in the first experiment random blocks should have created a significant advantage over alternation. With such an advantage not being observed, it is assumed the feature divider does not use the parameter of the last trial when the next trial is unpredictable possibly because the consequence of a wrong parameter might be high. In other words, the underlying mechanism is able to choose to work either preparing for a certain feature or not. This account suits with previous work that discovered a distinction between 'detecting a feature' and 'detecting a singleton' by Bacon and Egeth in 1994. In the second experiment, however, there was no advantage for predicted alternation over the random blocks. This might be because a feature divider theoretically can hold only one parameter at a time, so that replacing the old parameter with a new one results in the loss of the old parameter.

Naturally, there are alternative accounts of the expectation-repetition effect. One account states that the effect is caused from negative priming, which is the disadvantage of attending to an object or a feature that was formerly inhibited, as stated by Tipper in the year 1985. There are two arguments that can be offered against the negative-priming account.

First, expectation is an important component of the expectation-repetition effect. It isn't implausible that a feature divider would be significantly affected by voluntary control, as in the act of conscious expectation. However, voluntary control plays a minor role in the negative priming where the negative priming of a certain feature won't stop simply because of the lack of expectation. Secondly, it is believed that negative priming reflects post-perceptual factors as stated by Neill in 1997. Thus, negative priming shouldn't show up in a measurement of perceptual accuracy of brief displays.

A compelling conclusion could be drawn from Moore and Egeth in 1998 and several of the mentioned studies. The conclusion states that feature information does not result in an instantaneous improvement to the perception of a specific feature. There are several possible accounts for the phenomenon. Feature information can affect only by guiding spatial attention. In other words, spatial attention mediates feature-based attention. Location is what is called the 'master map' of visual attention. Search operations on a certain feature are facilitated by, and performed following to, the highlighting of all locations containing that feature. This notion is widely expressed in attention models, such as Cave's model in 1999, and the 1989 model of Wolfe, Cave, and Franze, as well as being supported experimentally, such as the experiments of Johnston and Pashler in 1990.

At first, the work made by Huang and Pashler appears to contradict the feature-based attention studies made by Farell and Pelli in 1993, Shih and Sperling in 1996,and Moore and Egeth in 1998. These studies suggest that the feature information cannot affect the perception in brief displays, but the present study shows the exact opposite. However, further analysis of the rationale and of the distinctions made between the methods in the previous and the current study that the result of Huang and Pashler indicate that their result in fact strengthens that their notion of 'spatial mediation of color-based attention'. According to the spatial mediation rationale of feature-based attention, the results of Moore and Egeth in 1998 should be interpreted in a matter where knowledge about a target feature must be translated into location information, which is then used to start the attention processing of other search dimensions.

An important distinction between the preset work and the three studies that have been mentioned is that subjects were required to report secondary features in the latter studies, whereas the study of Huang and Pashler required the subjects to report only location. In a study such as that of Moore and Egeth in 1998, the location map produced by the feature divider can be used by selective attention, but the effect of the feature divider cannot be observed because the usage is slow.

There are three conclusions made by the five experiments. First, the perception of feature cues can be improved by priming. The priming effect isn't triggered by expected repetition nor is it triggered by expected alternation alone, but it is triggered only by expected repetition. Secondly, the perceptual improvement cannot be fully explained by either target-feature enhancement or distractor-feature inhibition, nor by their summation. Finally, the expectation-repetition effect reported is different from the repetition priming effect. The repetition priming effect depends merely upon repetition instead of expectation, and appears to be exclusively post-perceptual. Meanwhile, the expectation-repetition effect requires both repetition and expectation, and appears to have perceptual components.

The conclusions given are very important in the case of the cryptid as it shows that there is a lack of perceptual improvement in nearly all cases of encounters. Nonetheless, psychology fits in with this piece of evidence.

When toddlers reach out in order to touch an object in front of them, they learn about texture, shape, and vision. The concept of body movement being necessary for vision was demonstrated with a pair of kittens in 1963.

Richard Held and Alan Hein, two researchers at MIT, placed a pair of kittens into a cylinder that is ringed in vertical stripes. Both of the kittens received visual input from moving around inside the cylinder. There was a critical difference, however, in the experience of the two kittens. The first kitten was walking on its own accord, whereas the second kitten was riding in a gondola that was attached to a central axis. Because of the setup, both kittens saw the stripes moving at the same time and speed. If vision was just about photons the eyes, then the visual systems of the two kittens should identically develop. Surprisingly, the results show that only the kitten that was walking on its own accord developed normal vision, whereas the kitten riding in the gondola never learned to see probably.

Vision isn't just photons that can be interpreted by the visual cortex with ease. Instead, vision is a whole body experience. The signals that arrive into the brain can only be made sense at by training, which requires cross-referencing the signals with the information from the actions and sensory consequences of a person. It is the only way the brain can interpret the visual data.

Perception requires the brain to compare different streams of sensory data against one another. However, the issue of timing makes this type of comparison challenging. The brain of different speeds processes all streams of sensory data. An example is a sprinter at a racetrack. Between the gunfire and the time that sprinters get off the block, there is a sizeable gap of almost two tenths of a second. Athletes train to make the gap as small as possible, but biology imposes the fundamental limits where the brain has to register the sound, send signals to the motor cortex, and then down the spinal cord to the muscles of the body.

Experiments show that sprinters responded more slowly to light than a bang. The reason behind this is the speed of information processing. Visual data goes through more complex processing than auditory data. Thus, it takes a larger span of time for signals that carry flash information to work their way through the visual system than the time required for bang signals to work through the auditory system. While it took 190 milliseconds to respond to light, sprinters reacted in 84.2% of that time to a bang at a mere 160 milliseconds.

In most cases, the mere thirty milliseconds or so would activate the expected alternation. Thus, the illusion of Bigfoot and its relatives is formed. Even so, this is a far-fetched theory considering the absurd reaction time. Therefore, more evidence should be given in order to place a plausible theory.

Firstly, there are specific sensory processes. Specialized nerve cells that evolved to be excited by particular external stimuli, such as light or sounds, provide the sense experience. Also called afferent neurons, stimulated receptor cells set up a chain reaction that excites neighboring cells to create a neural pathway along 'connector neurons' to the central nervous system and the brain. Signals from the brain travel on similar pathways to efferent neurons, also called motor or effector neurons, in order to stimulate muscles and control bodily movement.

In the brain, electrochemical 'messages' from the sensory organs are experienced as sensation. Each sense sends a different type of signal, and different regions of the brain, which are analogues to the different sensory organs, process the signals. For example, the primary visual cortex receives information from the eyes. The information is then analyzed and interpreted by the neighboring visual association. To apply this to the several relatives of Bigfoot, the information must first be applied to cognitive psychology.

The term 'cognitive psychology' is currently associated with the approach to psychology that became predominant after World War II, focusing on the mental processes instead of behavior. However, psychologists had set out to study the manner that the brain works from the earliest days of psychology. Cognitive psychology formally emerged in the 1950's with the cognitive revolution. Cognitive revolution is a movement largely influenced by advances made in both the information and computer sciences.

A pioneer in the scientific study of memory, which belongs in cognitive psychology, was Hermann Ebbinghaus; a German psychologist who examined his own ability to remember lists of words and letters in the 19th century.

Hermann Ebbinghaus was born in the 24th of January, 1850, in Barmen, Germany. At the age of 17, Ebbinghaus started his education at the University of Bonn with history and philosophy. Later on, however, Ebbinghaus studied in Berlin and Halle. The Franco-Prussian War interrupted Ebbinghaus studies in 1870, where he joined the Prussian army at the age of 20. After the war, Ebbinghaus continued and completed his doctoral dissertation in philosophy at the university of Bonn in 1873. Between the years of 1873 and 1880, Ebbinghaus studied in England and France, and began to increasingly become interested in psychology.

Ebbinghaus lectured at the University of Berlin from 1880 to 1886. During this time, Ebbinghaus had a son named Julius in 1885, who grew up to become a well-known philosopher. After Ebbinghaus lectured at the University of Berlin, he became a professor in Berlin from 1886 to 1894. Finally, Ebbinghaus had a chair in Berlin and Halle in 1894. Ebbinghaus founded laboratories in Berlin and Breslau, as well as teaching in Halle. Ebbinghaus was also one of the founders of the '*Zeitschrift für Psychologie und Physiologie der Sinnesorgane*'. Some of his important publications are '*Über die Hartmannsche Philosophie des Unbewußten*' in 1873, two volumes of '*Grundziige der Psychologie*' in 1902, and '*Psychologie*' in 1907. Hermann Ebbinghaus died in 1909 in Breslau, Germany, which is located in what is now known as Poland, due to pneumonia, an infection that inflames air sacs in one or both lungs, which may have been filled with fluid.

From Ebbinghaus most well known observations, he identified patterns of how people learn and forget, and his discoveries laid the foundations for the current study of memory. However, the methods of Ebbinghaus were less influential than the discoveries. As there is no direct access to other people's minds, Ebbinghaus believed that introspection, the examination or observation of a person's own mental processes, is the only method to study mental processes. Other experimental psychologists viewed introspection as subjective and unscientific. These experimental psychologists went on to devise other experimental techniques to study cognitive processes as well as several different subjects.

One discovery Ebbinghaus made is that two thirds of the information is lost after a span of 24 hours. However, Ebbinghaus discovered that there are several methods to overcome the rapid and exponential 'forgetting curve'. Information sticks within the brain with repetition. Repetition also causes information to be easier to retrieve from memory.

Results from Ebbinghaus's experiments show distinct patterns of memory and forgetfulness. Together with a rapid onset and slow decline that marks the forgetting curve, Ebbinghaus noticed a similar learning curve that is associated with memorization.

Bluma Zeigarnik, a Russian psychologist, discovered another distinctive feature of memorization while she was watching her local café. Waiters were able to remember exactly what a person ordered if asked when the order hasn't been paid yet. However, the waiters found it to be much more difficult to recall the order after the order has been paid. Memory of a completed transaction was no longer important, and thus, they have been put aside for new orders. In later experiments, Zeigarnik discovered than an unfinished task is generally better remembered than a task that has been completed.

There are two types of memory storage: short-term memory and long-term memory. Short-term memory stores information for a span of a few seconds with a limited capacity, whereas long-term memory can store an unrestricted amount of information in an indeterminate span of time. Short-term memory deals with the information that is required to immediately be used. Meanwhile, long-term memory deals with the information that is required for future use. Most psychologists recognize the dual-store model of memory. However, there is some disagreement on the exact roles of short-term memory and long-term memory, their connection, and whether the two are indeed separate systems.

The concept of memory storage produced the idea that there is a physical region in the brain that stores memories, or at least, that memory is stored in a specific region of the brain. However, in the 1940's, Karl Lashley discovered that memory is evenly distributed across the brain, as opposed to what has previously been thought. Donald Hebb, Lashley's colleague, went on to describe how people learn in terms of neural connections.

Every action or experience triggers a distinct pattern of neural connections. Hebb explained that if an action or experience is repeated, neural connections would strengthen and become hardwired into the brain as assemblies of cells. Thus, people learn by making associations between different assemblies.

The mind organizes and interprets the sensory information that is sent to it from the sensory organs instead of experiencing the sensory information directly as sensation. This is the cognitive process called perception. In order to interact with the external world, the mind makes sense of what is seen, heard, touched, smelled, and tasted, as well as the capability to distinguish between the important and irrelevant information.

Perception is the ability to distinguish the foreground from the background, and identify objects and their position. Most of the time, this occurs unconsciously. Some psychologists, especially those that are in the Gestalt movement, believe that perception is hardwired into the brain. In other words, humans are programmed to organize information into meaningful forms. Other psychologists, however, believe that perception is learned from experience.

Perception directly affects memory, both of which are important factors in the scientific explanation of the many cultural views on the primate cryptid. As the mind organizes and interprets the sensory information, the concept of the external world is capable to vary with an extremely minor margin. Afterwards the image is stored in the long-term memory. However, memory gradually fades with time. The reason itself is not time, however, but other memories. There are a finite number of neurons, all of which are required to multitask. Each neuron engages in different tasks at different times. The neurons operate in an active matrix of changing relationships, and substantial requirements are continuously applied on them. Therefore, memories are muddied as new memories appear. Each new event needs to establish new relationships among a finite number of neurons.

The pioneering work of Professor Elizabeth Loftus at the University of California, Irvine, conveyed clues to the malleability of memory. Professor Loftus devised an experiment where she asked volunteers to watch videos of car crashes. Afterwards, Loftus asked the volunteers several questions to test what the volunteers could recall. The questions that were asked influenced the given answers. Loftus explained:

"When I asked how fast were the cars going when they hit each other, versus

how fast were the cars going when they smashed into each other, witnesses give different estimates of speed. They thought the cars were going faster when I used the word '<u>smashed</u>'."

From there, Loftus set out to figure out if it is possible to insert absolutely incorrect memories. Loftus recruited a range of participants, and her team contacted the participants' families to retrieve information about events that occurred in the participants' past. Using the information, the researches assembled four stories about the participants' childhood experience. Three stories were true. On the other hand, the fourth story had plausible information, yet it never occurred to any of the participants. The fourth story states that the participant was lost as a child in a shopping mall and reuniting with a parent with the help of a kind elderly person.

In a succession of interviews, the participants were told the four stories. More than 25% of the participants claimed that they are able to remember the incident in the shopping mall without it actually occurring. Loftus explained the phenomenon:

"They may start to remember a little bit about it. But when they come back a week later, they're starting to remember more. Maybe they'll talk about the older woman, who rescued them."

As time progressed, more details began to emerge from the false memory, such as the reaction of their parent, what the participants had with them, and the clothing of the elderly person. This applies to many 'Bigfoot experts' that supposedly had 'absolute evidence' for the existence of the cryptid, only to belong to a bear, a known species of primate, a deer, or any other known species of mammal. Thus, it can be concluded that it is possible to insert false memories into the brain. People would embrace and elaborate the false memories, unknowingly fabricating fantasy into their own identity.

All people are susceptible to memory manipulation. For example, when Elizabeth Loftus was a child, her mother had drowned in a swimming pool. After several years while on a conversation, Elizabeth's relative had told Elizabeth that she had been the person to discover her mother's body in the pool. That came as a shock to Elizabeth as she didn't know that. Elizabeth didn't believe that statement, but she described the following:

"I went home from that birthday and I started to think: maybe I did. I started to think about other things that I did remember - like when the firemen came, they gave me oxygen. Maybe I needed the oxygen 'cause I was so upset I found the body?"

Soon, Elizabeth was able to visualize her mother in the swimming pool. However, her relative had later on called to say that he had made a mistake.

It wasn't Elizabeth who discovered the body, but rather her aunt. That is how Elizabeth experienced the feeling to possess false memory, which was richly detailed and deeply felt.

The past of people is not an accurate record. Rather, the past is a reconstruction that can sometimes border on mythology. Not all details are accurate when life memories are reviewed. Some details come from stories while others were filled in with what was thought to what must have happened.

Assuming that the estimation of around 25% that have received false memory due to the details being filled in by either the stories of the cryptid that they have heard from others or from filling in what was thought what must have happened, then further estimations must be made. It is nearly impossible to have the complete data of eyewitnesses throughout the Earth, thus further data must be calculated. Taking in the fact that there are 3,400-eyewitness accounts in both the United States of America and Canada, with 20% more eyewitnesses in each of these two countries than any other country, then one country would have around 1,360 eyewitnesses at most. Multiplying this number by the 65 countries that the five mentioned versions of the cryptid live in, with the exception of the North American Bigfoot, then the amount of eyewitnesses is equal to 88,400. Adding this to the number of eyewitnesses of North America, then the number of eyewitnesses of the six versions of the cryptid is more or less than 91,800.

With a very rough estimation of the amount of eyewitnesses, then the next step would be calculating the number of people that saw the creature due to false memory. As the base number of that number is 25%, there are 22,950 people who saw it due to false memory. Yet, that would not explain the rest of the 75% of people, which is equal to 68,850, spotting the creature. The next part of elimination involves honesty. Studies show that there is a 60% chance of a person to lie, as opposed to 40% of telling the truth. If the studies are indeed accurate, then there is a mere 0.05809% chance of all 68,850 people to tell the truth. Out of the 68,850 people to claim to have spotted the creature, 27,540 of those people are deemed to be trustworthy. With 70% of eyewitnesses disregarded as a hoax, there is yet even more evidence to support the fact that the ape cryptids are a hoax.

At the study made from Liqiang Huang and Harold Pashler, a heightened vision and reaction time are not dependent on expectation alone. However, they are dependent on both expectation and repetition. Therefore, for the eyewitnesses to fully see the creature in detail, they would have spotted it before and expected the creature to return. However, this would cause a problem. If true, then there are both a pattern to the movement of the creature and hotspots to easily spot the cryptid. Yet, for the two centuries in which the cryptid has been popularized, the reason behind the lack of scientific evidence proving the existence is the lack of both the pattern and the hotspots. Thus, the capability of spotting the same cryptid at least twice with expectation is very rare, if not impossible.

If so, then the creature spotted would be vague. There are currently three major pieces of evidence that haven't been discussed yet for the existence of Bigfoot: Bigfoot corpses.

Nearly all corpses, although extremely rare, are easy to analyze. The build and lack of hair of corpses typically show a type of bear with alopecia. Alopecia is a type of disease that causes hair to fall out in small patches. This disease develops when the immune system attacks the hair follicles, which then results in hair loss. There are two types of alopecia: alopecia totalis, which is the hair loss on the scalp, and alopecia universalis, which is the hair loss on the entire body.

Alopecia is an autoimmune disease. An autoimmune disease develops when the immune system confuses the healthy cells with alien material. Generally, the immune system defends the body against invaders, such as viruses and bacteria. With alopecia, however, the immune systems attack the hair follicles instead. Hair follicles are the structures from which hairs grow. The follicles would eventually shrink and stop the production of hair, which leads to hair loss.

There are several cases of animals with alopecia, and two of the most famous cases are that of bears and dogs. Thus, it is clear that the most likely answer to the identity of the first two corpses is that they are black bears with alopecia.

In one case, however, the story of a Bigfoot corpse went viral. In January 2014, Rick Dyer released images of the body of the alleged Bigfoot. Dyer claimed to have shot and killed the Bigfoot in a wooded region on the northwest side near Loop 1604 and Highway 151 in early September of 2012. Dyer stated the following after reclaiming the body from his investors:

"*I have been worried for so long; I have been put off for so long and finally we went up to Washington (state) and we got the body. Every test that you can possibly imagine was performed on this body, from DNA tests to 3D optical scans to body scans. It is the real deal. It's Bigfoot and Bigfoot's here and I shot it and now I'm proving it to the world.*"

Despite a history of several Bigfoot hoaxes in the past, with a Bigfoot that Dyer claimed to kill in 2008 being a rubber ape costume, Dyer insisted that his kill was genuine. However, this was proven to be fake. A few months after the supposed Bigfoot body was revealed at a press conference, it was identified to be nothing more than a rubber ape costume.

With even corpses being a hoax, the people that are persuaded that the corpses, and thus the existence of the cryptid, are able to persuade skeptics using psychological factors.

Firstly, there is a form of human automatic action that is suitably revealed in an experiment made by the Harvard social psychologist Ellen Langer. A well-known principle of human behavior states that asking a person to do a favor, it would be more successful if a reason were provided. It is of the human nature to like having a reason for their actions. Langer demonstrated the fact by asking a small favor of people waiting in line to use a copying machine in a library. It is shown that 94% of those asked let Langer to skip ahead of them in line when she told him that she had five pages and that she's in a rush. Meanwhile, only 60% of those asked complied when Langer did not provide her reasoning. While it might be appear that the difference between the two requests was the fact that Langer was in a hurry, a third type of request made by Langer proved otherwise. It was not the entire reasoning that caused the difference, but the use of 'because' that made the difference. The result of the third experiment shows that 93% of people agreed.

People are not expected to recognize and analyze all of the details in even an entire day. The human brain doesn't have the time, energy, or capacity to do so. Instead, people are required to use their stereotypes very often, and the rules of thumb to classify objects according to a handful of key features and then mindlessly respond when a trigger feature is present.

Sometimes the behavior that unrolls won't be appropriate for the current circumstance, because stereotypes and trigger features don't work all the time. However, their imperfection must be accepted, as there isn't any other choice. From all indications, people would rely on the stereotypes and trigger features to a more considerable extent in the future. As the stimuli that saturate the lives of people continue to grow to be more intricate and variable, people will have to depend increasingly on their shortcuts to handle all of them. The renowned British philosopher Alfred North Whitehead recognized the inescapable quality of modern when he asserted that people could perform without thinking about civilization advances by extending the number of operations.

There are several components shared by most of the weapons of automatic influence. They are the nearly mechanical process by which the power within the weapons can be activated, the consequent exploitability of the power by any person that knows to trigger the weapons, and the way that the weapons of automatic influence lend their force to the people who use the weapons.

The process of the weapons lending their force to the person using them is much more advanced than providing obvious weaponry to be used by a person to bludgeon another person into submission. With proper implementation, the exploiter would not need to strain a muscle to get their way. The only necessary effort is to activate the great supplies of influence that exist in the situation and direct them towards the target. In that sense, the approach is similar to that of jujitsu, a form of Japanese martial arts. A person that employs jujitsu would make use of their own strength minimally against their opponent. Instead, the person would utilize the power intrinsic in natural principles such as gravity, leverage, momentum, and inertia. If the person knows how and where to engage the action of the principles, then they can effortlessly defeat a person that is physically stronger. So it is for the exploiters of the weapons of automatic influence that exist naturally. The exploiters are able to commission the power of the weapons for use against their targets while exerting little force themselves. The last feature of the process allows the exploiters a colossal benefit, which is the ability to manipulate the targets without it appearing as manipulation. Even the targets themselves tend to spot their obedience as determined by the action of natural forces.

This is one method in which 'Bigfoot experts' and social media manage to persuade people that they have spotted the cryptid, hence triggering other illusions in the mind such as false memory and the expectation-repetition effect.

There is a principle in human perception, called the contrast principle, that affects the way that people see the difference between two separate things that are presented one after another. In other words, if the second item is fairly different from the first, people will tend to see the second object as more different than it actually is. For example, if a person were to lift a light object and then lift a heavy object, then they would estimate that the second object to be heavier than they would if they had lifted the object before holding the light one. The contrast principle is well established in the field of psychophysics and applies to all types of perception other than weight. If a person were to talk to an attractive woman and are followed by an unattractive woman thereafter, then the second woman would seem to be less attractive to the person than she actually is.

Studies done at Arizona State and Montana State universities about the contrast principle show that people might be less satisfied with the physical attractiveness of their own lowers because of the way that the popular media bombards people with several examples of unrealistically attractive models. In a study, college students rated an image of an average-looking member of their opposite sex as less attractive if the student had first looked through ads in some of the popular magazines. In another study, the males that reside in college dormitories rated the photo of a potentially blind date. Those that rated the photo while watching an episode of the TV series *Charlie's Angels* viewed the blind date as a less attractive female than the males who rated the female while watching a different show. Seemingly, it was the uncommon beauty of the female stars that made the blind date to seem less attractive.

The advantage of the principle is not only that the principle works, but also that the principle is virtually undetectable. Those who are able to employ the principle can cash in on its influence without any appearance of having structured the situation to be in their favor. An example of this is the Patterson-Gimlin film. While many are persuaded that the claims of Roger Patterson and Robert 'Bob' Gimlin are genuine, it should be noted that an advantage of the stated principle is that it is virtually undetectable, and do not clearly show to have structured the incident in their favor. However, Roger Patterson was confirmed to require evidence for the creature because the creature itself fascinated him, and that he was writing a book himself. Thus, the pair has unintentionally triggered one of the major methods to psychologically persuade people, which thereafter caused them to have unwanted fame and bad reputation amongst those that have not been persuaded by Roger Patterson and Robert Gimlin.

Several years ago, a university professor tried an experiment. The professor sent Christmas cards to a number of strangers. Although the professor expected some reaction, he was still amazed when holiday cards that are addressed to him came streaming back from the strangers. The great majority of the strangers who had returned a card never inquired into the identity of the professor. While the study itself might seem small, the study shows the action of one of the most potent of the weapons of influence, the rule for reciprocation. The rule states that people should try to repay, in kind, what another person has provided them. By the virtue of the reciprocity rule, people are obligated to the future repayment of favors, gifts, invitations, etc.

The impressive aspect of the rule for reciprocation and the sense of obligation is its pervasiveness in the entirety of human culture. This is so widespread that after an intensive study, sociologists such as Alvin Gouldner are able to report the fact that there aren't human societies that don't subscribe to the rule. Within each society, it seems pervasive and permeates exchanges of every type. The renowned archaeologist Richard Leakey attributes the essence of what makes a human to the reciprocity system:

"We are human because our ancestors learned to share their food and their skills in an honored network of obligation."

Cultural anthropologists Lionel Tiger and Robin Fox view the web of indebtedness as a distinctive adaptive mechanism of humans, which allows for the division of labor, the exchange of various forms of goods, the exchange of different services, and the creation of a bunch of interdependencies that unite individuals together into highly efficient units.

A study made by a pair of Canadian psychologists detected something fascinating about people at a racetrack: Immediately after placing a bet, people are more confident of the chances of their horses winning the race than they were before placing the bet. Although it is puzzling at first, the reason for the change has to do with a common weapon of social influence. Similar to the other weapons of influence, the weapon lies deep within a person, directing their actions with silent power. Simply, the weapon is the nearly obsessive desire to be consistent with what the person has already done. Once a person has made a choice or taken a stand, they would encounter personal and interpersonal pressures to consistently behave with the commitment. Those pressures would cause the person to respond in methods that justify their earlier decision. For example, thirty seconds before the bettors in the racetrack experiment put down their money; they had been tentative and uncertain. Thirty seconds after the deed, however, the bettors were much more optimistic and self-assured. The act of making a final decision, such as the act of buying a ticket, had been the critical factor. Once a stand had been taken, the need for consistency pressured the people to bring what they had felt and believed into line with what they had already done. The people persuaded themselves that they had made the correct choice and, without a doubt, feel better about it.

Psychologists have long understood the power of the consistency principle to direct human action. Prominent theorists such as Leon Festinger, Fritz Hieder, and Theodore Newcomb have viewed the dire for consistency as a central motivator of the behavior of humans. The tendency to be consistent is, without a doubt, powerful enough to compel a person to do what they ordinarily would not be willing to do.

An example comes from an experiment made by the psychologist Thomas Moriarty. Thomas Moriarty staged thefts on a beach in New York City to see if witnesses would risk personal harm to halt the theft. In the study, a research accomplice would put a beach blanket down at around 1.5 meters, or 5 feet, from the blanket of a randomly chosen individual, which thus became the experimental subject. After a few minutes on the blanket spent relaxing and listening to music from a portable radio, the accomplice would get up and leave the blanket in order to stroll down the beach. A few minutes afterwards, a second researcher, who pretended to be a thief, would approach, grab the radio, and try to hurry away with the radio. Under normal conditions, subjects were very reluctant, and only 20% of the subjects put themselves in harm's way to challenge the thief. However, another experiment was made with the only difference being that the accomplice would ask the subject to watch their things, and all subjects agreed. In this experiment, which is propelled by the rule for consistency, 95% of the subjects ran after and stopped the thief. Sales and motivation consultant Cavett Robert captures the principle in his guidance to sales trainees:

"*Since 95% of the people are imitators and only 5% initiators, people are persuaded more by the actions of others than by any proof we can offer.*"

Researchers have also employed procedures based on the principle of social proof. One psychologist in particular, named Albert Bandura, has led the way to develop such procedures in order to elimination of undesirable behavior. Bandura and his colleagues have shown how people that suffer from phobias can be rid of the extreme fears in a simple fashion. For example, an early study chose several nursery-school-age children because of their fear of dogs. The children merely watched a little boy happily playing with a dog daily for twenty minutes. The exhibition produced marked changes in the reactions of the fearful children that 67% of the children were willing to climb into a playpen with a dog after four days. Moreover, the researchers found that the improvement had no evaporated in a month when the researchers tested the children's fear levels again one month after the original experiment.

An important practical discovery was made in a second study of children who were unusually afraid of dogs. To reduce their fears, it was unnecessary to provide live demonstrations of a child playing with a dog. Film clips showed the same effect as the previous experiment. The most effective type of clips was the clip depicting a variety of other children interacting with their dogs. Apparently, the principle of social proof works best when the proof is provided by the actions of a lot of other people.

The powerful effect of filmed examples in changing the behavior of children could be used as therapy for several problems. Research made by the psychologist Robert O'Connor on socially withdrawn preschool children shows some striking evidence for therapy. O'Connor was worried that a long-term pattern of isolation was forming that would create persistent complexities in social comfort and adjustment through adulthood. In an attempt to reverse the pattern, O'Connor made a film that contained 11 different scenes in a nursery-school setting. Each scene began by showing a different isolated child watching some continuous social activity and then joining the activity to everyone's excitement. O'Connor then selected a group of the most severely withdrawn children from four preschools and them the film that he made. The isolated children immediately began to interact with their peers at a level equal to that of the normal children in the schools. O'Connor found that six weeks thereafter, the students that have viewed his film were now leading their schools in their amount of social activity.

This applies to many Bigfoot experts and the majority of supporters in both newspapers and social media. Not only do the clips supposedly showing Bigfoot have the same effect as the clips of the children playing with the dogs, but so does the conclusion. The conclusion states that the social proof is most efficient when it is provided by the actions of many other people. With tens of thousands of eyewitnesses and actions made by the media, the efficiency of the social proof is skyrocketed.

While that would not make all of the remaining 30%, or 27,540 people, to be persuaded, the number of people that fall into the category of persuasion is outstanding. It has been concluded that 67% were influenced by vision while hearing influenced 75%. However, believing in the existence of the cryptid would require a combination of both. Thus, the original number of 91,800 must once again be given for the calculation.

Persuasion can affect several of the mentioned factors of the reason behind believing to have witnessed the cryptid. Specifically, persuasion would trigger the expectation-repetition effect as well as bending the original memory of the incident into false memory. This would link persuasion into illusion and perception. Thus, the effect of persuasion must be calculated on the 91,800 people instead of 27,540. The effect of persuasion can therefore be easily calculated as 142% of the 27,540 people. This would therefore lead to at least 39,107 people being directly affected by the several psychological methods of persuasion.

CONCLUSION

Bigfoot and its relatives have been stated for a long period of time, and have played a major role in the history of mankind. From cultural and spiritual beliefs to an increased amount of tourism and camping, these giant apes have brought us together. The cryptids have partially and indirectly caused an increased amount of respect and curiosity towards enthusiastic people. Globally, many people have turned to science, math, and nature during the last two centuries to find answers to the identity of the cryptids.

Even so, there has been a major split between the believers of the creature and the non-believers. Whether we believe in its existence or not, we still cannot deny that there have been countless theories about the many versions of Bigfoot. From the paranormal and the religious to the scientific and the hypothetical, there is no answer that would be considered to be definitive towards all of the people throughout the world.

The theory stated in this book is nothing more than the conclusion reached by the scientific and mathematical data given. To summarize the conclusion, Bigfoot is non-existent and is seen in many different manners, whether as a misidentified animal, a cultural legend, a practical joke, or any other lesser-known manners. From there, the legends grew by perceptual errors, psychological rewiring of the incident, effects of scientific persuasion, or simply computer effects and/or human actors. Until hard evidence of the cryptid appears to debunk this theory, this is the main conclusion given to the identity of the legends.

ACKNOWLEDGMENTS

→, V. (2018). Infographic shows 3,313 sightings of Bigfoot in 92 years. [online] Doubtful News. Available at: http://doubtfulnews.com/2013/09/Infographic-shows-3313-sightings-of-bigfoot-in-92-years/ [Accessed 2 Jan. 2018].

Adventure Nation Blog. (2018). Himalayan People - Life in the Himalayas. [online] Available at: http://www.adventurenation.com/blog/himalayan-people-life-in-the-himalayas/ [Accessed 2 Jan. 2018].

Anon, (2018). [online] Available at: https://www.researchgate.net/post/What_is_an_accurate_way_to_measure_surface_area_of_a_large_animal [Accessed 2 Jan. 2018].

Anon, (2018). [online] Available at: https://www.ewt.org.za/Reddata/pdf/Primates(4)/2016%20Mammal%20Red%20List_Cercopithecus%20albogularis_VU.pdf [Accessed 2 Jan. 2018].

Bbc.co.uk. (2018). BBC - GCSE Bitesize: Adaptations - cold climates. [online] Available at: http://www.bbc.co.uk/schools/gcsebitesize/science/aqa_pre_2011/evolution/organismsrev2.shtml [Accessed 2 Jan. 2018].

Center, M. (2018). Barometric Pressure and Migraines: What You Need to Know. [online] Blog.themigrainereliefcenter.com. Available at: http://blog.themigrainereliefcenter.com/barometric-pressure-and-migraines-what-you-need-to-know [Accessed 2 Jan. 2018].

Encyclopedia Britannica. (2018). Himalayas | History, Map, & Facts. [online] Available at: https://www.britannica.com/place/Himalayas [Accessed 2 Jan. 2018].

Faculty.ycp.edu. (2018). Cite a Website - Cite This For Me. [online] Available at: http://faculty.ycp.edu/~brehnber/BOA%20lectures%20&%20labs/Surface%20Area%20and%20Volume%20of%20Animal%20Bodies%20HO.pdf [Accessed 2 Jan. 2018].

Geolsoc.org.uk. (2018). The Geological Society. [online] Available at: https://www.geolsoc.org.uk/Plate-Tectonics/Chap3-Plate-Margins/Convergent/Continental-Collision [Accessed 2 Jan. 2018].

Joshuastevens.net. (2018). Joshua Stevens - Visual Communicator. [online] Available at: http://www.joshuastevens.net/ [Accessed 2 Jan. 2018].

Keratin.com. (2018). hair color - biochemistry. [online] Available at: http://www.keratin.com/as/as002.shtml [Accessed 2 Jan. 2018].

Legal Beagle. (2018). Why Don't People Report Crimes to the Police?. [online] Available at: https://legalbeagle.com/5733254-dont-people-report-crimes-police.html [Accessed 2 Jan. 2018].

Listverse. (2018). Top 10 Hybrid Animals - Listverse. [online] Available at: http://listverse.com/2011/05/02/top-10-hybrid-animals/ [Accessed 2 Jan. 2018].

Mentalfloss.com. (2018). 60% of People Can't Go 10 Minutes Without Lying. [online] Available at: http://mentalfloss.com/article/30609/60-people-cant-go-10-minutes-without-lying [Accessed 2 Jan. 2018].

prezi.com. (2018). Hominidae vs Pongidae. [online] Available at: https://prezi.com/9b4zfmblysog/hominidae-vs-pongidae/?webgl=0 [Accessed 2 Jan. 2018].

ScienceDaily. (2018). Why an ice age occurs every 100,000 years: Climate and feedback effects explained. [online] Available at: https://www.sciencedaily.com/releases/2013/08/130807134127.htm [Accessed 2 Jan. 2018].

sciencenordic.com. (2018). Grizzly-polar bear hybrids spotted in Canadian Arctic. [online] Available at: http://sciencenordic.com/grizzly-polar-bear-hybrids-spotted-canadian-arctic [Accessed 2 Jan. 2018].

Smithsonian Insider. (2018). Why did Neanderthals go extinct?. [online] Available at: http://insider.si.edu/2015/08/why-did-neanderthals-go-extinct/ [Accessed 2 Jan. 2018].

www.prehistoric-wildlife.com, D. (2018). Prehistoric Wildlife. [online] Prehistoric-wildlife.com. Available at: http://www.prehistoric-wildlife.com [Accessed 2 Jan. 2018].

Zihlman, A. and Underwood, C. (2013). Locomotor Anatomy and Behavior of Patas Monkeys (Erythrocebus patas) with Comparison to Vervet Monkeys (Cercopithecus aethiops). Anatomy Research International, 2013, pp.1-11.

(n.d.). Retrieved January 2, 2018, from http://www.blueplanetbiomes.org/rainforest.htm

Shmoop Editorial Team. (2008, November 11). Spanish Colonization. Retrieved January 2, 2018, from https://www.shmoop.com/spanish-colonization/

M. (1999, June 09). The Amazon Rainforest. Retrieved January 2, 2018, from https://rainforests.mongabay.com/amazon/

Gold in South Africa. (n.d.). Retrieved January 2, 2018, from http://www.goldinsouthafrica.com/pdfs/Chapter_2.pdf

Cobbing, A. S. (2017, February 11). South Africa. Retrieved January 2, 2018, from https://www.britannica.com/place/South-Africa

National Geographic. (n.d.). Retrieved January 2, 2018, from https://www.nationalgeographic.com/

WWF - Building a future in which humans live in harmony with nature. (n.d.). Retrieved January 2, 2018, from http://wwf.panda.org/

Glossotherium robustum. (n.d.). Retrieved January 2, 2018, from https://prehistoric-fauna.com/Glossotherium-robustum

Snopes.com. (2011, August 25). Retrieved January 2, 2018, from https://www.snopes.com/

Cialdini, R. B., PH.D. (n.d.). INFLUENCE The Psychology of Persuasion. Retrieved January 2, 2018, from http://elibrary.bsu.az/books_400/N_232.pdf

Gorilla Evolution. (2017, February 01). Retrieved January 2, 2018, from http://www.gorillas-world.com/gorilla-evolution/

Nature's Giants: 27 of the world's biggest critters. (n.d.). Retrieved January 2, 2018, from http://www.foxnews.com/science/slideshow/2014/05/06/natures-giants-27-worlds-biggest-critters.html#/slide/eastern-lowland-gorilla

Alopecia Areata: Causes, Symptoms, and Diagnosis. (n.d.). Retrieved January 2, 2018, from https://www.healthline.com/health/alopecia-areata

H, S. G. (n.d.). Retrieved January 2, 2018, from http://stochastikon.no-ip.org:8080/encyclopedia/en/ebbinghausHermann.pdf

Ecology. (n.d.). McDougal Littell Science.

Geography: A children's encyclopedia. (n.d.). DK.

Hart-Davis, A. (Ed.). (n.d.). History: From the Ancient to the Modern World. DK.

Muir, H. (2013). Science in seconds: 200 key concepts explained in an instant. New York: Quercus.

The Natural History Nook: TheUltimate Guide To Everything on Earth. (2010). London: DK.

ABOUT THE AUTHOR

Alwaleed S. Alghanim is an aspiring paleobiologist. Alwaleed has been mainly self-taught, but has also learned from his experience with scientists and professors, including the Oxford professor Sharif Siam. He has also been in the 'Mawhiba', a program aimed at the gifted students of the Middle East. His current focus is to teach and inspire other young people.

For contact: **Alwaleedalghanim01@gmail.com**

www.ingramcontent.com/pod-product-compliance
Lightning Source LLC
Chambersburg PA
CBHW031629210526
45464CB00004B/1822